Introduction to Chaos and Coherence

T0139325

Introduction to Chaos and Coherence

Jan Frøyland

Department of Physics
University of Oslo

Institute of Physics Publishing

Bristol and Philadelphia

British Library Cataloguing-in-Publication Data

A catalogue record for this book is available from the British Library.

ISBN 0-7503-0194-5 (hbk)
     0-7503-0195-3 (pbk)

Library of Congress Cataloging-in-Publication Data are available

First printed 1992
Reprinted 1994 (hbk + pbk)

Published by Institute of Physics Publishing, wholly owned by
The Institute of Physics, London

Institute of Physics Publishing, Techno House, Redcliffe Way,
Bristol BS1 6NX, UK

US Editorial Office: Institute of Physics Publishing, The Public Ledger Building,
Suite 1035, Independence Square, Philadelphia, PA 19106, USA

Typeset by TEX at Institute of Physics Publishing
Printed in the UK by J W Arrowsmith Ltd, Bristol

CONTENTS

PREFACE

Chaotic phenomena flourish in nature. They often originate in systems whose components are governed by simple laws, but whose overall behaviour is very complex. The concept of fractals has developed from being something exotic, known by a only few mathematicians, to a topic explained to the layman on the TV, and the study of complex, nonlinear systems has become a frontier in many areas of science.

This short book aims to give an elementary introduction to the theory of chaotic systems and demonstrate how chaos and coherence are interwoven in some of the simplest models exhibiting deterministic chaos. This is part of the theory somewhat more formally called dynamical systems theory.

The origin of this book was a set of lecture notes for a short course in dynamical systems theory given at the University of Oslo for the first time in 1985. In the time available as many as possible of the concepts of the theory were developed. However, no attempt has been made to go into any depth on any particular subject, and many important topics have been totally omitted.

It is hoped that people who are engaged in making models will find the book useful and, perhaps even more, that those who have to rely on other people's models can get some good hints on how to ask pertinent questions—it is important to be aware of the fact that chaotic behaviour implies the impossibility of long time predictions, and that such behaviour is normal in nonlinear models.

Extensive use has been made of Lyapunov exponents as a tool that can make the exploration of a particular model less time consuming in terms of human effort. A chapter about time series analysis contains methods of analysis that have not yet become standard. In particular, people who make models intended to describe some time series can find a prescription of how to find the scaling dimension which gives the absolute minimum of variables needed in the model.

There are no references in the text, for which I apologize very much. The reference section contains only a selection of books. The list is by no means intended to be complete. By and large, it contains the books used in writing this book, and therefore also serves as an acknowledgement of my debt to the authors of these books. The reader who wants to search for the original literature is recommended the book by Hao Bai-lin which contains a collection of many of the most significant papers in the field and in addition contains a reference section listing several hundred papers. Also, the books by Mandelbrot contain extensive lists of references. Hao mainly refers to publications by non-mathematicians, while Mandelbrot mainly refers to mathematicians.

The notation in general does not need much explanation except that in higher dimensional maps the space index is an upper index with the possibility of confusion with an exponent. The advantage is that there will never arise any doubt about what is the space and what is the time index.

In the past I have greatly benefited from conversations with Predrag Cvitanovic and also Robert H G Helleman. I am particularly grateful to my former student Kai T Hansen who has given me permission to use colour pictures in addition to figures and other unpublished material from his thesis. He has also read parts of the manuscript and made several useful suggestions. I would also like to thank Helge Sandvei for valuable assistance.

Jan Frøyland
September 1991

1

INTRODUCTION

Many of the fundamental properties of classical dynamical systems were discovered in the last century by people like Poincaré and Birkhoff. The advent of the new quantum physics distracted the attention of physicists and chemists from these problems for half a century. The revival of the field is closely related to the availability of modern computers to perform the necessary numerical experiments. In particular, Feigenbaum's discovery of universality gave the field a big push forward. The implications are that many totally unrelated systems have in common features both qualitatively and, even more amazingly, also numerically.

Any system that develops in time in a non-trivial manner may be considered a dynamical system, but to make it somewhat clearer what *dynamical systems theory* is about it is perhaps appropriate to call it *nonlinear dynamical systems theory* since all the interesting effects that we shall encounter are closely related to the nonlinearity of the systems under consideration. In particular, we shall consider systems that possess elements of so-called *deterministic chaos* and how chaos is reached from non-chaotic states as controlling parameters are varied.

Dynamical systems are normally regulated by parameters. When the parameters change, so do the properties of the system. In particular, the stability of a system may be investigated by considering the results of small disturbances. If the disturbances die with time the system is stable, and if the disturbances grow the system is unstable. At some points in the space of parameters some of these properties may change discontinuously as a function of one or more of the parameters. When linear systems lose stability it is usually obvious why. It may be because the controlling parameters have changed exponential decay into exponential growth or because some boundary condition is 'violated'. However, many nonlinear dynamical systems lose stability for no obvious reason, in which case more or less dramatic changes of dynamical patterns take place. This kind of phenomenon—unknown to the linear theory—is known as a *bifurcation*. Investigating bifurcations in some detail is one of the major subjects of this book.

Before we proceed it may help to make a rough classification of various dynamical systems and their basic characteristics. The classification is given in table 1.1. The concept *phase space* is somewhat loosely taken to mean the entire space spanned by the minimum number of dynamical, i.e.

Table 1.1 Basic properties of classes of dynamical systems

Class	Phase space dimension	Variable types	Time
Partial diff. equations	infinite	continuous	continuous
Ordinary diff. equations	finite	continuous	continuous
Iterative maps	finite	continuous	discrete
Cellular automata	finite	discrete	discrete

time-dependent, variables of the problem. We shall mainly be concerned with low-dimensional maps and differential equations, but we shall also have a brief look at the simplest properties of cellular automata. The systems under consideration are mostly *deterministic*, i.e. they are systems where the next time step is always exactly predictable. Exceptions to this are the *autoregressive models* where the updating in time has a *stochastic* component.

We shall not deal with other interesting aspects of nonlinear dynamics, such as solitons and related topics.

2

FRACTALS

Exercises in traditional classical mechanics usually consist of considering a set of integrable equations of motion. This means that the equations of motion may be completely separated into as many *independent* equations as there are degrees of freedom. The equations are then said to be an integrable set of equations even though it may not be possible to express the solutions in terms of elementary functions. In ideal—fortunate—cases the integrations may be performed and the results—the solutions—be presented as functions of time and initial conditions. Unfortunately, this lucky state of affairs is in some sense abnormal. In realistic models of systems one will frequently find that the equations of motion are not completely integrable, meaning that there are fewer independent separation constants than the number of degrees of freedom. One might make an analogy with the real numbers where the irrational numbers outnumber the rational ones. The non-integrability usually comes about because of nonlinearities or because of the presence of external forces. If the system is non-integrable the orbit that the system traces out in phase space will normally be *chaotic*. The most important (necessary but not sufficient) property of a chaotic orbit is that it never returns to a point previously visited and that it is also not approaching a periodic orbit.

Although it may not be possible—even in principle—to express the motion of a system in phase space as a function of time and initial conditions, the desire to find some quantity or object that does not change with time is very strong. If the motion is periodic one may give coordinates of points on the *orbit*—the phase space trajectory—but only rarely is this of much help, and if the orbit is *chaotic* this is quite useless unless the orbit is subjected to some kind of analysis. Even if the motion is chaotic the orbit may nevertheless be bounded in all directions in phase space and attracted to geometrical objects called *strange attractors* with strange and unfamiliar properties. In particular, a strange attractor does not change with time.

Strange attractors are members of a family of mathematically defined objects named *fractals*. The term fractal was invented by Mandelbrot who gives the following explanation of the word:

> Fractal comes from the Latin adjective fractus, which has the same roots as fraction and fragment and means 'irregular or fragmented'. It is related to *frangere* which means 'to break'.

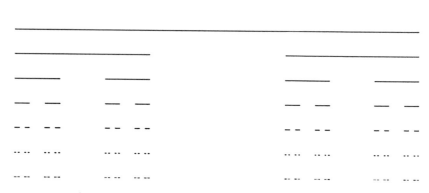

Figure 2.1 Construction of the Cantor set for $b = 1/3$.

Typically *it is not sensible to assign an integer dimension to a fractal.*
However, there are several alternative ways to arrive at a non-integer dimension of a given fractal. This we shall return to later. The notion of
fractals is useful in many circumstances both in describing phenomena in
nature (clouds, mountains, rivers, trees, coast lines, etc) and to give a frame
of reference for strange mathematical concepts previously avoided by the
non-mathematician.

2.1 A Cantor set

As a first example of a fractal we shall consider a simple Cantor set. Take
a straight line of unit length. Remove from its middle a section of length
b where $0 < b < 1$. Now there remain two pieces, each of length $(1 - b)/2$.
Remove from the middle of each of these pieces a bit of relative length b,
i.e. a piece of length $b(1 - b)/2$. What remains after having performed
this operation an infinity of times is a particular fractal called a Cantor
set. Having repeated this operation n times there are 2^n line segments of a
total length $(1-b)^n$, so that the length of each segment is $l_n = (1-b)^n/2^n$.
One way to define a fractal dimension, D, that may be applied in this case
is through the equation

$$L = \text{constant} \times l^{1-D} . \tag{2.1}$$

Here L is the total 'length' of the set as measured with a measuring stick
of length l and the *constant* is independent of l. When all possible length
scales are applied, and one finds that D is independent of l, then equation
(2.1) may be considered a defining equation for the fractal dimension D.
The reader should convince her/himself that the definition (2.1) is in agreement with the common sense dimensions of a point or a non-fractal curve.

In the case of a point, only one measuring stick is needed to cover the set, and hence $L = l$ independent of l, and it follows from equation (2.1) that $D_{point} = 0$. In the case of a straight continuous line $L = $ constant independent of l, and consequently $D_{line} = 1$.

Returning to our case of the Cantor set, the *constant* may be eliminated by considering two consecutive generations of the operation of removing middle line segments. Let L_n be the total length of the remaining line segments after n operations, and l_n be the length of each remaining piece and also the length of the measuring stick in question. One then arrives at

$$\ln\left(\frac{L_n}{L_{n+1}}\right) = (1 - D)\ln\left(\frac{l_n}{l_{n+1}}\right). \tag{2.2}$$

Evidently $L_{n+1} = (1 - b)L_n$ and $l_{n+1} = (1 - b)l_n/2$. Inserting this into equation (2.2) gives the fractal dimension

$$D = \frac{\ln 2}{\ln\left(2/(1 - b)\right)}. \tag{2.3}$$

Using $b = \frac{1}{3}$ (see figure 2.1) gives the fractal dimension of this Cantor set $D = 0.6309\ldots$. Notice that the total length of the remaining line segments approaches zero as n goes to infinity, implying that the topological dimension of the set is zero. Still this Cantor set has something in common also with the line. This is expressed in the fact that $0 < D < 1$.

The Cantor set has another very interesting property. It is *self-similar*. This means that one may take any small section of the set and enlarge it by *rescaling* it a factor (in this case 3) any number of times and by a suitable translation make it cover a part of the original set exactly.

2.2 The Koch triadic island

As a second example of a fractal we shall consider the Koch triadic island. It is constructed by starting with a regular triangle with sides of unit length. Each side is divided into three equally long pieces. Then the middle piece is replaced by two pieces of length $\frac{1}{3}$ in the fashion shown in figure 2.2. This process is continued *ad infinitum*. In a manner completely equivalent to the previous case we get for the total length of the 'coast' of the 'island':

$$L_n = 3(4/3)^n \tag{2.4}$$

and

$$l_n = (1/3)^n. \tag{2.5}$$

Using l_n as a measuring stick we get from equation (2.2)

$$D = \ln 4/\ln 3. \tag{2.6}$$

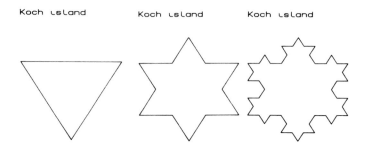

Figure 2.2 The three first generations in the construction of Koch's triadic island.

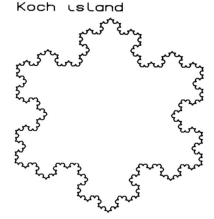

Koch ιslαnd

Figure 2.3 Fifth generation construction of Koch's triadic island.

By now we have examples of the definition of the fractal dimension being capable of producing results both less than and bigger than one. At this point a word of warning may be in order: the knowledge of the fractal dimension of an object does not give very precise information of what the object looks like, just as there is an infinity of lines of topological dimension one. The construction above is such that at any level the curve may be stretched into a closed curve. Therefore the topological dimension of the Koch curve is one, its fractal dimension is 1.2618 ... and it is embedded in a two-dimensional space.

From the preceding example one may get the impression that the fractal dimension is necessarily non-integer. This is not so, as may be seen from the next example. Follow the construction above for the Koch curve, but in addition, at each step, remove one of the middle two pieces. This evidently leaves the total length equal to 3, independent of generation number. However, the line is no longer continuous, and the topological dimension of this construction is zero, while the fractal dimension is exactly one.

In the three examples mentioned here the fractal dimension has been greater than the topological dimension and less than the dimension of the space in which the fractal is embedded. Indeed we may take this to be a necessary condition any object must satisfy to be a genuine fractal.

2.3 Fractal dimensions

In the previous examples we used the connection between length and dimension given by equation (2.1) to find the fractal dimension. In more general cases this definition does not work. The fractal dimension most often referred to is probably the *Hausdorff dimension*. However, its definition is such that it is clumsy to calculate on computers, and we shall instead use the dimension called the *capacity* which may be computed using the *box counting* algorithm.

Assume the space to be metric (i.e. one can define sensible distances) and of integer dimension d. In this space we may cover the fractal with d-dimensional regular boxes all of equal size. Their sizes are regulated by one dimensionless length or scale parameter, l. This can always be achieved by choosing non-dimensionless basic units so that l becomes just a scaling variable. If, for instance, the space is two dimensional, say temperature against time, the box size corresponding to the length parameter l may simply be l °C $\times$ l s. One then counts the number $N(l)$ of boxes that are not empty. $N(l)$ is connected to the capacity or box counting dimension, d_c, by

$$N(l) = \text{constant} \times l^{-d_c}. \qquad (2.7)$$

By plotting $\ln(N(l))$ against $\ln(l)$ for many values of l, a proper fractal yields a straight line from which the dimension is calculated. This definition of a fractal dimension is frequently used in computer experiments. However, in spaces of dimensions greater than three this procedure often runs into difficulties both with computer memory space and time. There is another difficulty about this definition applied in practice, and it is that the result may not be independent of the shape of the boxes. In the above example we might get a different result if the sizes of the boxes were l °C $\times$ l min.

In figure 2.4 we show another example of a self-similar fractal construction. This is the triangular Sierpinski gasket. Its construction is evident from the figure. The dimension can easily be calculated using the box counting algorithm. The boxes are triangles, and starting with a box of size one, only one box is needed to cover the figure. It takes three boxes of size $\frac{1}{2}$, and nine boxes of size $\frac{1}{4}$ and so on, to cover the non-empty parts of the figure. The capacity dimension then becomes $d_{\text{Sierpinski}} = \ln 3 / \ln 2$.

The defining equation (2.7) may give results that are counter-intuitive. Consider for instance the set of all rational numbers in the interval $\langle 0, 1 \rangle$.

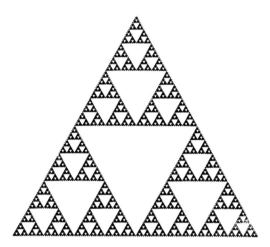

Figure 2.4 Triangular Sierpinski gasket with fractal dimension $d = \ln 3 / \ln 2$.

Since there are always rational numbers in any interval of finite length, the capacity dimension of the set is one in spite of the fact that it has measure zero.

In a number of cases the capacity dimension and the Hausdorff dimension are the same. In general $d_c \geq d_H$, where d_H is the Hausdorff dimension. It may seem unreasonable not to take into account the density of points inside each box, or for a dynamical system how much time is spent inside each box. This may be taken into account by what is known as the *Kolmogorov entropy* and its generalization: the *Procaccia–Henschel dimension*.

More complicated fractals may be considered to be unions of fractal subsets. Such a *multifractal* may consist of an infinity of fractal subsets, each with its own dimension. This kind of object is better characterized by a continuum of dimensions.

3

THE LOGISTIC MAP

Consider the iterative map

$$x_{n+1} = f(x_n) \qquad (3.1)$$

where the function $f(x)$ is used to 'predict' the next x with x_n as a starting point. This may be considered to be a discrete time series of a variable, x, measured at regular, finite time intervals and where n is the time label. The system is *deterministic* since a given x value always results in the same, predictable successor.

To avoid unnecessary indices we shall sometimes use the notation $x \leftarrow f(x)$ instead of equations like (3.1).

Iterative maps of this type have received a great deal of attention. Apart from the fact that many people find these maps fascinating, there also exist physical reasons for studying discrete maps. For instance, they can mimic many important features of realistic nonlinear dynamical systems and they can be used to reduce the complexity and dimensionality of phase space. How this may be done is explained to some extent in section 3.17 about Poincaré maps and return maps.

As a prelude to the *logistic map*, which is what we get in the case where the function $f(x)$ of equation (3.1) is a parabola, we shall take a brief look at a rather trivial but important case, namely the *linear map*.

3.1 The linear map

Let $f(x)$ be just the straight line:

$$f(x) = Ax + B \qquad (3.2)$$

i.e.

$$x_{n+1} = Ax_n + B. \qquad (3.3)$$

Introduce now the *scaling* and *translation* transformation

$$x_n = kx'_n + d. \qquad (3.4)$$

The scaling parameter k and the translation parameter d may be chosen for our convenience. Inserting equation (3.4) into equation (3.3) gives

$$kx'_{n+1} + d = kAx'_n + Ad + B. \qquad (3.5)$$

Choosing $d = Ad + B$, which is always possible provided $A \neq 1$, one finds

$$x'_{n+1} = Ax'_n. \tag{3.6}$$

The parameter b of equation (3.3) is therefore not interesting since it may (almost) always be transformed away. Also the scaling transformation was of no use to us in this case since it drops out of equation (3.5) when d is chosen so that the constant term is zero.

Dropping the prime, equation (3.6) has the following solution:

$$x_n = x_0 A^n \tag{3.7}$$

where x_0 is any number. In particular, $x_0 = 0$ is the *fixed point* of the map so the orbit is just $x_n = x_0$ independent of n. For $|A| > 1$ the solution represents an exponentially growing amplitude, while $|A| < 1$ represents an exponentially damped motion. In the last case $x_n \to 0$ as $n \to \infty$. This means that the orbit is attracted to the fixed point which is then said to be an *attractor* for $-1 < A < 1$. Obviously the fixed point is stable since any orbit that starts a little way away from the fixed point will gradually approach it. For $|A| > 1$ the motion is forever divergent, and in this case the fixed point is unstable and is therefore said to be a *repellor*. No matter how close the orbit may be to the repellor initially, it can always be found arbitrarily far away after a sufficiently long time.

The case $A = 1$ may be taken to represent undamped oscillations observed at time intervals equal to the time strobing interval. Also $A = -1$ represents an oscillating system but with period 2 since it takes two iterations to get back to the starting point. This particular case will be important in order to understand the phenomenon of *period doubling bifurcations*.

3.2 Definition of the logistic map. Scaling and translation transformations

Returning to a more general $f(x)$, we may consider a series expansion around a fixed point and keep terms up to order two in x:

$$f(x) = A + Bx + Cx^2. \tag{3.8}$$

This $f(x)$ appears to be a three-parameter function, but in reality it is only a one-parameter function because of the freedom to make scale and translation transformations as in equation (3.4). Using this transformation we obtain

$$x_{n+1} = (A - d + Bd + Cd^2)/k + (B + 2Cd)x_n + Ckx_n^2. \tag{3.9}$$

The scaling parameter k may now be used to fix the coefficient of the second-order term to anything we like (except zero, of course). Similarly, fixing the translation parameter d to a convenient value effectively leaves only one parameter. This freedom to choose the form of $f(x)$ has given rise to several essentially equivalent expressions. More or less common are the following:

$$f(x) = 2cx + 2x^2 \qquad (3.10)$$
$$f(x) = 1 - \mu x^2 \qquad (3.11)$$
$$f(x) = 4\lambda x(1 - x) \qquad (3.12)$$
$$f(x) = c + x^2 \qquad (3.13)$$
$$f(x) = b(1 - 2x^2). \qquad (3.14)$$

Any of these equations together with equation (3.1) may be considered the definition of the logistic map.

The relations between the various parameters can easily be found from equation (3.9). For graphical illustrational purposes the form (3.12) is perhaps the most practical, while in many calculations the form (3.13) involves the least amount of writing.

As an example, let us find the transformation that connects the parameter λ of equation (3.12) with the parameter c of equation (3.13). From equation (3.8) we have $A = 0$, $B = 4\lambda$ and $C = -4\lambda$. Inserting this into equation (3.9) to yield equation (3.13) we find

$$(-d + 4\lambda d - 4\lambda d^2)/k = c$$
$$4\lambda - 8\lambda d = 0$$
$$-4\lambda k = 1.$$

This gives $k = -1/4\lambda$ and $d = \frac{1}{2}$ and

$$c = 2\lambda(1 - 2\lambda). \qquad (3.15)$$

3.3 The fixed points and their stability

The fixed points of the map are those that satisfy

$$x = f(x). \qquad (3.16)$$

Using equation (3.12) one finds two solutions:

$$x_1 = 0 \qquad (3.17)$$

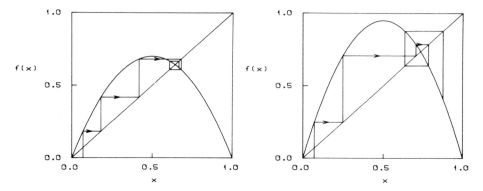

Figure 3.1 Graphical illustration of the iterative procedure with $\lambda = 0.70$ to the left and $\lambda = 0.95$ to the right.

and

$$x_2 = 1 - 1/(4\lambda). \qquad (3.18)$$

In figure 3.1 two constructions are made of orbits that start out near the fixed point $x = 0$. In the case $\lambda = 0.7$ the second fixed point is an attractor, while in the case $\lambda = 0.95$, the fixed point is a repellor. The motion starts out to be divergent in a way similar to what we found in the linear case for $A > 1$, since the orbit moves away from the nearby fixed point, but whereas in that case the motion continued to be divergent forever, the nonlinear term now prevents that from happening even in the case when both the fixed points are unstable.

We may make a more formal investigation of the stability of the fixed points: let the fixed point be defined by equation (3.16) and start the orbit a small distance away from x. Define the small quantities ε_n by

$$x_n = x + \varepsilon_n. \qquad (3.19)$$

Inserting this and expanding to first order in the small quantities gives

$$x + \varepsilon_{n+1} = f(x + \varepsilon_n) \approx f(x) + f'(x)\varepsilon_n \qquad (3.20)$$

or

$$\varepsilon_{n+1} \approx f'(x)\varepsilon_n \qquad (3.21)$$

which is nothing but the linear map of equation (3.6). As a consequence we must have the *stability condition*

$$|f'(x)| < 1 \qquad (3.22)$$

fulfilled for the fixed points to be stable. In graphical terms this means that a fixed point is unstable whenever the absolute value of the slope of the function $y = f(x)$ is greater than one at the intersection point(s) with the line $y = x$. In our case

$$f'(x) = 4\lambda - 8\lambda x. \tag{3.23}$$

Inserting the values of the fixed points as functions of λ, one finds stability for

$$-\tfrac{1}{4} < \lambda < \tfrac{1}{4} \quad \text{for } x = x_1 \tag{3.24}$$

and

$$\tfrac{1}{4} < \lambda < \tfrac{3}{4} \quad \text{for } x = x_2. \tag{3.25}$$

We shall confine our interest to the λ interval $\langle\tfrac{1}{4}, 1]$ where x_1 is always unstable.

If we consider $f'(x_2) = -4\lambda + 2$, where x_2 is the second fixed point, we see that it changes monotonically from 1 at $\lambda = \tfrac{1}{4}$ to -1 at $\lambda = \tfrac{3}{4}$. This has some very specific consequences that we shall return to. At the moment we shall only note that somewhere in the stability region, i.e. at $\lambda = -\tfrac{1}{2}$ there must be a parameter point where $f'(x_2)$ is zero. At this point the fixed point is said to be *superstable*. We know from the linear case that the normal convergence towards the attractor is exponential, but in the case of superstability the convergence is much faster. This can be seen directly since $\lambda = \tfrac{1}{2}$ is one of the very rare cases where it is possible to obtain an exact expression for any orbit. The result is

$$x_n = \left[1 - (1 - 2x_0)^{2^n}\right] / 2. \tag{3.26}$$

Here x_0 is any number in the interval $\langle 0, 1\rangle$. The relation (3.26) can be verified by direct insertion. Obviously this converges extremely rapidly.

We learned previously that the stable fixed point loses its stability at $\lambda = \tfrac{3}{4}$, but it still continues to be a fixed point, but at $\lambda = \tfrac{3}{4}$ it makes the transition from being an attractor to becoming a repellor. Nevertheless, its influence is very important also for $\lambda \in \langle\tfrac{3}{4}, 1]$.

To find out more about what happens at $\lambda = \tfrac{3}{4}$ we may perform what has become a standard numerical experiment: first we fix a starting value of λ then we pick an initial x_0 at random in the interval $\langle 0, 1\rangle$ and let the computer run for a while (100 iterations will do for most purposes) to allow transients to die out. Then the plotting of the x_n values starts. How many values one plots of course depends on the resolution of the plotting device—200 to 300 x values at each λ value often gives reasonably good pictures. Slowly increasing λ gives a picture like figure 3.2. At $\lambda = \tfrac{3}{4}$ the single line representing the stable period one splits into two through a *pitch-fork bifurcation*. Another word for the same thing is a *period doubling bifurcation*. The concept of bifurcations is essential to the whole theory of dynamical systems, and we shall encounter other types besides period doubling bifurcations.

Let us make the stability analysis somewhat more complete at $\lambda = \frac{3}{4}$. Since $f'(x_2(\lambda = \frac{3}{4})) = -1$, it follows from equation (3.21) that the small disturbance is making a period two oscillation. Consequently we now understand why there is a period doubling bifurcation at $\lambda = \frac{3}{4}$.

A period N orbit is an orbit that returns to the starting point after N iterations. The reason why the fixed point—the period one orbit—is stressed so much is that it is the prototype of all periodic orbits. This can be seen by the following consideration: the equation to determine a (there may be many) period N orbit is

$$x = f(f(\cdots(f(x))\cdots)) \equiv f^N(x) \tag{3.27}$$

where there are N fs. However, this is nothing but the fixed point equation for the new function

$$F(x) \equiv f^N(x). \tag{3.28}$$

Thus it is a general result that we have a period doubling bifurcation whenever $F'(x) = -1$, and a *tangent bifurcation* when $F'(x) = 1$. The stable period three orbit first appears in the chaotic region as a result of a tangent bifurcation with $N = 3$. We shall return to this in section 3.13.

3.4 Period two

To find period two orbits we need to solve the equations

$$
\begin{aligned}
x_1 &= 4\lambda x_2(1 - x_2) \\
x_2 &= 4\lambda x_1(1 - x_1).
\end{aligned} \tag{3.29}
$$

Subtracting and dividing out the period one solutions ($x_1 = x_2$) one gets

$$x_1 + x_2 = 1 + 1/(4\lambda) \tag{3.30}$$

and

$$x_{1,2} = (4\lambda + 1 \pm \sqrt{(4\lambda + 1)(4\lambda - 3)})/(8\lambda). \tag{3.31}$$

Obviously this orbit exists only for $\lambda > \frac{3}{4}$.

Now we may carry out a stability analysis similar to what we did before. Defining $F(x) = f(f(x))$ then x_1 or x_2 of equation (3.31) is just a fixed point of the function $F(x)$. Thus

$$
\begin{aligned}
F'(x_{1,2}) = f'(x_1)f'(x_2) &= (4\lambda)^2(1 - 2x_1)(1 - 2x_2) \\
&= 1 - (4\lambda + 1)(4\lambda - 3).
\end{aligned} \tag{3.32}
$$

When $\lambda = \frac{3}{4}$, $F'(x_{1,2}) = 1$ as expected at the point where the orbit first occurs. Putting $F'(x_{1,2}) = -1$ to find the point where the orbit bifurcates

into a period four orbit, one finds $\lambda = (1 + \sqrt{6})/4 = 0.8624\ldots$. We find
the point where the orbit is superstable by putting $F'(x_{1,2}) = 0$ giving
$\lambda = (1 + \sqrt{5})/4 = 0.8090\ldots$.

The period one orbit is stable in an interval in λ of length $\frac{3}{4} - \frac{1}{4} = \frac{1}{2}$,
while the period two orbit is stable in an interval of length $(\sqrt{6}-2)/4$. The
ratio between these two numbers is $4.449\ldots$. A similar ratio can be formed
between the lengths from the superstable points of the period one and the
period two orbits to their bifurcation points respectively. The result is
$\sqrt{6} + \sqrt{5} = 4.685\ldots$. The significance of these numbers will be explained
in the next section.

3.5 The period doubling route to chaos. Feigenbaum's constants

In the previous section we have seen that the period one orbit bifurcates into
a period two orbit which again bifurcates into a period four orbit. From
figure 3.2 we see that this sequence of successive bifurcations continues.
The poor resolution prevents us from seeing very many, but in fact there is
an infinity of period doubling bifurcations occurring at shorter and shorter
intervals. After n bifurcations the length of the period is 2^n, so after
an infinity of bifurcations the period is infinitely long, i.e. there is no
periodicity any more and the attractor of the system has become *aperiodic*
or *chaotic*. The point where this happens, λ_∞, is an *accumulation point* of
the period doubling sequence.

Suppose we designate by λ_n the point where the period 2^n orbit bifurcates. One may then form the ratios

$$\delta_n = \frac{\lambda_n - \lambda_{n-1}}{\lambda_{n+1} - \lambda_n}. \tag{3.33}$$

Feigenbaum made some remarkable observations concerning this sequence.
The first one is that the δ_n series very quickly converges to a constant
$\delta = 4.669\,20\ldots$. This means that the approach to λ_∞ is geometric. An
alternative definition of δ is to use the distances between the superstable
points which are easier to determine numerically than the distances between
bifurcations. This converges even faster than the previously defined δ_n
sequence to the same number δ.

Using these facts one can make an estimate of the position of λ_∞ by
assuming the series to be exactly geometric from the $(n-1)$th bifurcation
on. One finds

$$\lambda_\infty \approx \lambda_n + \frac{\lambda_n - \lambda_{n-1}}{\delta - 1}. \tag{3.34}$$

Using $n = 2$ and the approximate value $\delta = \sqrt{6} + \sqrt{5}$ from the previous section we obtain $\lambda_\infty \approx 0.892\,862\ldots$ as compared to a more precise
numerical result $\lambda_\infty \approx 0.892\,864\ldots$.

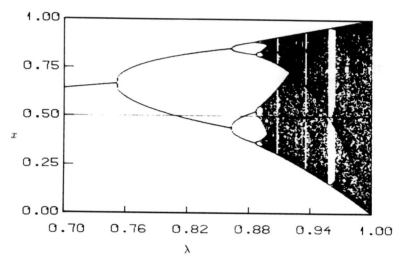

Figure 3.2 The result of the numerical experiment: the felled bifurcation tree which shows the attractor plotted against the parameter λ. The line $x = \frac{1}{2}$ represents the critical point.

One can also define another ratio by measuring the size of the opening of one fork compared to the opening of the next generation of forks. This is illustrated in figure 3.3. The openings are measured at the point where the orbits are superstable. One then forms the ratios $\alpha_n = -\varepsilon_n/\varepsilon_{n+1}$. This sequence too converges quickly to a constant $\alpha = 2.5029\ldots$. Since this number is smaller than δ, the forks will look more and more like little vertical bars as λ_∞ is approached.

Feigenbaum discovered that both δ and α are *universal* numbers for period doubling cascades. In this connection that means that these numbers are independent of the particular form of $f(x)$ provided $f(x)$ is a smooth single hump function with continuous first derivative, and the second derivative is different from zero at the extremum.

From the result of the numerical experiment one might think that the period 2^n orbit does not exist byond λ_n, but just as we have seen analytically for the period one orbit, the orbit continues to be there all the way up to $\lambda = 1$, but now as a period 2^n repelling orbit and it will therefore not be visible in the type of computer experiment leading to figure 3.2 where only attractors may be seen. However, it is easy to find algorithms to locate unstable low period orbits also. Note that for all $n > 2$ there exist other stable and unstable period 2^n orbits in the region $\lambda_\infty < \lambda < 1$ than those that are born in period doubling bifurcations at $\lambda < \lambda_\infty$.

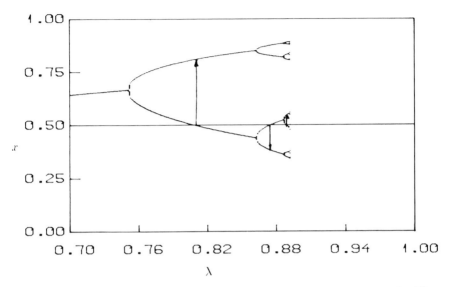

Figure 3.3 Illustration of how the openings of the forks are measured. The horizontal line represents the critical point.

3.6 Chaos and strange attractors

Beyond λ_∞ there exist chaotic orbits. The hallmarks of a chaotic orbit are

- it is not periodic,
- it does not have a periodic orbit as a limiting set,
- it has sensitive dependence on initial conditions.

The last requirement means that two orbits that start out close to another will become more and more separated—at least initially—as time runs. In the region $\lambda < \lambda_\infty$ we have seen that there exist stable periodic orbits. Periodic orbits that start out away from the stable periodic orbit will have it as a limiting set as time goes to infinity. In addition, a chaotic orbit will either have a limiting set called a *strange attractor* or it will move on the strange attractor itself. The degree of chaos may be measured by Lyapunov exponents which will be treated in sections 3.10, 5.3 and 9.3.

We call the parameter interval $\lambda_\infty \leq \lambda \leq 1$ the chaotic region, but that does not mean that there are only chaotic orbits in this region. In fact there is an infinity of *windows* where periodic attractors exist, and a periodic attractor may be found arbitrarily close to any strange attractor.

3.7 The critical point and its iterates

The point in phase space where $f'(x) = 0$ is called the *critical point*. With our particular parametrization the critical point is at $x = \frac{1}{2}$ independent of λ. (For many purposes it is more practical to transform away the first-order term in $f(x)$ so that the critical point is at $x = 0$, as in equation (3.13).) Let $f^n(x)$ be defined as in equation (3.27) and consider a period n orbit $x_{n+1} = x_1$. Then by the chain rule

$$f^{n'}(x_1) = f'(x_n)f'(x_{n-1}) \ldots f'(x_1). \tag{3.35}$$

Thus, whenever the critical point is on the orbit it means that $f^{n'}(\frac{1}{2}) = 0$ which implies that not only is the orbit stable, but it is superstable. In figure 3.4 a plot is made of $f^n(\frac{1}{2})$ as a function of λ for $n = 1, 2, 3, 4$ and 5. The first two iterates are

$$f^1(\tfrac{1}{2}) = \lambda \tag{3.36}$$

and

$$f^2(\tfrac{1}{2}) = 4\lambda^2(1 - \lambda). \tag{3.37}$$

Consider the interval spanned by these two values at a given λ. Obviously no iterate may be bigger than the iterate of the critical point, so $x = \lambda$ is an upper limit for the iterative prosess. It is somewhat less obvious that the second iterate must be the lower limit. Let us use a construction like the one we used to get figure 3.1 and start at a very unfavourable point near the unstable fixed point $x = 0$. Evidently the iterates will be growing till the critical point is passed. Because of the monotonic decrease of $f(x)$ after the critical point is passed, no result can be smaller than the second iterate of the critical point.

Figure 3.4 reveals that there is just one stable period two orbit, but there are two period four orbits. One is the one we know already. It was born in the bifurcation of the period two orbit. The second occurs quite close to $\lambda = 1$ and is responsible for a window in the chaotic region. There is one period three orbit with an associated window, and we also see that there must be three period five windows. The difference between the orbits of the same periodicity is that, arranged in increasing order, the phase space points of various orbits are not visited in the same order.

Another very characteristic feature of figure 3.4 is that there is a point where all iterates higher than two meet. This point is evidently also a fixed point since repeated use of the mapping function gives only the input value as a result. A frequently used phrase in this connection is to say that there is a *collision with an unstable fixed point* or that the orbit is captured on an unstable fixed point. Similarly the orbits at parameter points where the nth and the $(n + k)$th iterates of the critical point cross, are captured on

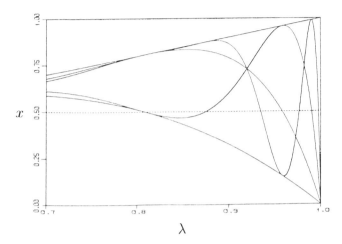

Figure 3.4 The five first iterates of the critical point as functions of λ. The first iterate is the line on top. For $n > 1$ the curve for the $(n+1)$th iterate is above the curve for the nth iterate near $\lambda = 1$.

unstable period k orbits. At the same time these crossing points are points where *inverse bifurcations* take place and their number is so great that they completely dominate the behaviour of the map in the chaotic region.

In figure 3.1, only the period three window is clearly visible, but in a blow up in figure 3.5 the leftmost of the period five windows is also seen. In that same figure also two period six orbits are seen, one as a result of the first period doubling bifurcation of the period three orbit.

It is easy to see that the nth iterate of the critical point is a polynomial in λ of order $2^n - 1$, and as n increases this becomes a function performing an extremely rapidly growing number of oscillations and crossings of the line representing the critical point. This means that there must be an infinity of periodic windows in the chaotic region. In fact one may wonder if there is anything else but periodic windows. However, it can be proven that when the total length of all periodic windows is added together there is still space left for the truly chaotic orbits, but a periodic orbit may be found arbitrarily close to any chaotic orbit. Note that in computer experiments the rounding off errors act as a source of noise which effectively puts a limit to how long periodic orbits can be seen.

On the other hand, a digital computer can represent only a finite number of points on the interval, and consequently sooner or later the points on the orbit must start to repeat.

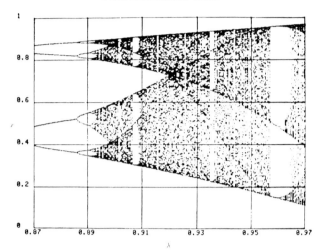

Figure 3.5 A blowup of figure 3.2.

3.8 Self-similarity, scaling and universality

From figure 3.2 one sees that the two forks of the period four orbit look very much like one fork of the period two except for the difference in scale and a slight distortion of the forks. The distortion is smallest for the fork that is crossed by the line representing the critical point. If the lowest period four fork in figure 3.2 is scaled up a factor δ in the λ direction, flipped around $x = \frac{1}{2}$ and streched a factor $-\alpha$ in the x direction, it will very nearly cover the period two fork. If this process is continued for orbits closer and closer to λ_∞, the error made will be relatively smaller and smaller and in the end will become zero. For this reason the hierarchy of bifurcations accumulating at λ_∞ is said to be *self-similar.*

Suppose we make a translation transformation so that the critical point is at $x = 0$. Assume further that we are at the accumulation point. Consider the 2^n times iterate of the point $x/(-\alpha)^n$. As n increases this becomes a point closer and closer to the critical point. The reader may find it intuitively reasonable that when $n \to \infty$ all memory of the original $f(x)$ is gone and the only thing that matters is what the function looks like at the critical point. Thus it can be shown that the function

$$g(x) = \lim_{n \to \infty} (-\alpha)^n f^{2^n}(x/(-\alpha)^n) \qquad (3.38)$$

exists and is independent of $f(x)$ for a large class of functions. The function $g(x)$ satisfies the *universality equation*

$$g(x) = -\alpha g(g(x/\alpha)). \qquad (3.39)$$

When one fixes $g(0) = 1$ the *functional equation* (3.39) uniquely determines

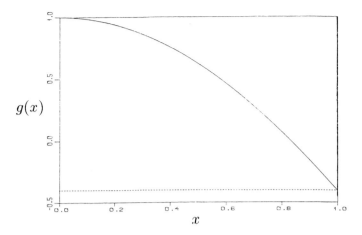

Figure 3.6 The function $g(x)$. The broken line is at $-1/\alpha$. See equation (3.40). For increasing values of x $g(x)$ oscillates more and more wildly.

$g(x)$ which accordingly must be totally independent of the original $f(x)$. Inserting $x = 0$ into equation (3.39) gives

$$g(1) = -1/\alpha. \tag{3.40}$$

Equation (3.39) is an equivalent of a fixed 'point' equation in *function space*. This may become clearer if we discretize x so that $g(x) \rightarrow g_i = g(x_i)$. We then get, instead of equation (3.39),

$$g_i = -\alpha g_l \tag{3.41}$$

where $x_l \approx g_k$ and $x_k \approx x_i/\alpha$. In the stable region of the period one orbit of the logistic map we could start at an arbitrary point and the orbit would converge to the fixed point. In a similar manner we may start with an (almost) arbitrary set of points g_i, use the rule (3.41), and the result will in the end be an approximation to $g(x)$ provided we fix $g_0 = 1$. The function $g(x)$ is shown in figure 3.6. However, this figure was made by choosing $n = 4$ in equation (3.38). This approximation to $g(x)$ is far better than the resolution of the graphical device. We shall not enter into any discussion of the *universality theory*; we only note that the constant δ is not determined by $g(x)$ itself; instead it is related to the stability of $g(x)$.

3.9 Reversed bifurcations. Crisis

In section 3.5 we learned that the first and second iterates of the critical point constituted the outer limitations of the iterative process after a possible transient period. By continuing the arguments leading to this conclusion one can find that the outer limitation must be the union of the intervals $\langle f^1(\frac{1}{2}), f^3(\frac{1}{2})\rangle$ and $\langle f^2(\frac{1}{2}), f^4(\frac{1}{2})\rangle$. For $\lambda > \lambda'_1$, where λ'_1 is the point where $f^3(\frac{1}{2}) = f^4(\frac{1}{2})$, the two intervals overlap, but as λ'_1 is passed from above the two intervals split. This phenomenon is called a *reversed bifurcation* or a *band splitting bifurcation*. Since further iterations of the fourth iterate of the critical point will only reproduce itself, this point must be a fixed point. Thus the condition for the inverse bifurcation to happen can be formulated in an equivalent way by saying that the first reversed bifurcation of *the main series* takes place where the unstable fixed point 'hits' the third iterate of the critical point. (It cannot hit the second iterate.) By definition all higher iterates are hit at the same point.

It is quite simple to calculate exactly the position of the first reversed bifurcation of the main series. The problem is simplest to solve using the parametrization (3.13).

The critical point is now at $x = 0$, and the third iterate of the critical point becomes

$$f^3(0) = c + (c + c^2)^2. \tag{3.42}$$

The fixed points are given by

$$x = \tfrac{1}{2} \pm \sqrt{\tfrac{1}{4} - c}. \tag{3.43}$$

The first reversed bifurcation of the main series takes place where the third iterate of the critical point hits the unstable fixed point. Setting the two expressions equal to another, rearranging and squaring, one finds

$$[c + (c + c^2)^2]^2 - (c + c^2)^2 = 0. \tag{3.44}$$

Factorizing and dividing out an uninteresting factor $c^4(2 + c)$ one is left with the following equation to solve:

$$c^3 + 2c^2 + 2c + 2 = 0. \tag{3.45}$$

This equation has only one real solution

$$c'_1 = \tfrac{1}{3}\left(\sqrt[3]{-17 + 3\sqrt{33}} - \sqrt[3]{17 + 3\sqrt{33}} - 2\right) \approx -1.543\,68. \tag{3.46}$$

Solving equation (3.15) to express λ in terms of c we find $\lambda'_1 \approx 0.919\,643$.

Moving in the direction of λ_∞ there is a new reversed bifurcation at the point λ'_2 where the unstable period two hits the sixth iterate (on the lower branch of figure 3.5) and the seventh iterate (top branch of figure 3.5). Continuing further in the direction of λ_∞ we can form a sequence of reversed bifurcation points which also accumulates at λ_∞, and whose approach to λ_∞ is geometrical with an associated constant δ which is precisely *the same* as we found for the period doubling bifurcation sequence!

Plate 1 Phase diagram for the circle map (chapter 4) produced by depicting the Lyapunov exponent in colours in the parameter range $-0.5 < \Omega < 0.5$ and $0.0 < K < 8.0$. The colour coding is: red: stable orbit. Light red means a small negative LE, dark red means large negative LE. Green/yellow is quasiperiodicity, and blue means positive LE, i.e. chaos. Compare to figures 4.1 and 4.5.

Plate 2 Phase diagram for the two-dimensional linearly coupled logistic map, section 6.3. The horizontal axis is c, and the vertical axis is d. $d = 0$ approximately in the middle of the picture. Red: stable orbit, green: quasiperiodicity, blue, yellow and white: various degrees of chaos. Compare to figure 6.10 for positive d.

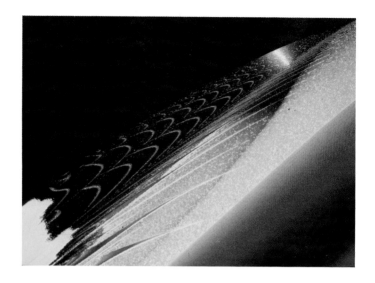

Plate 3 An enlargement of plate 2 in the parameter range $-0.661 < c < -0.322$ and $0.198 < d < 0.440$ showing Arnol'd tongues and a chaotic region with overlapping resonances. See figure 6.10.

Plate 4 An enlargement of plate 2 showing Hopf bifurcations with Arnol'd tongues along the border line between greeen and red in the parameter range $-0.7344 < c < -0.6700$ and $0.025 < d < 0.11$. See figure 6.10.

Plate 5   An enlargement of plate 4 in the parameter range $-0.697 < c < -0.680$ and $0.0802 < d < 0.093$. An Arnol'd tongue (red) undergoes yet another Hopf bifurcation.

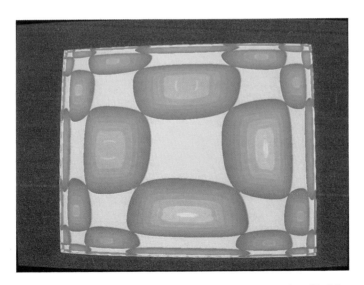

Plate 6   Basin of attraction at $c = -0.6$ and $d = 0.03$ in region II of figure 6.10. Orbits starting inside one of the ovals have the out-of-phase period two orbit as the attractor. Orbits starting at points outside the ovals have the in-phase period two orbit as the attractor.

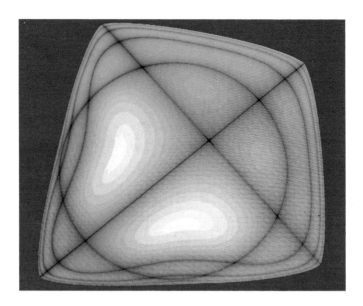

Plate 7 The structure of the basin of attraction at $c = -0.43$ and $d = 0.13$ in region I of figure 6.10. There is only one attractor. Dark colour indicates a long time before the orbit comes close to the attractor.

Plate 8 Two interwoven fractal basins of attraction at $c = -0.382\,723$ and $d = 0.3996$ in region III of figure 6.10. Two coexisting attractors, both period 22. Blue indicates one basin, and green/white indicates the other basin of attraction.

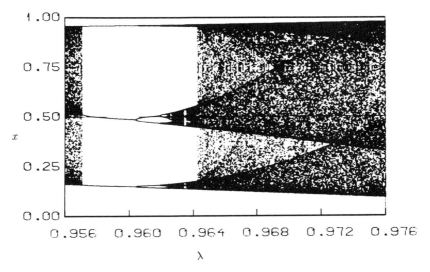

Figure 3.7 Magnification of figure 3.2 around the period three window.

The self-similarity of the chaotic region is now quite apparent: take the section of figure 3.2 lying between λ_∞ and $\lambda = 1$. Diminish it by a factor δ in the λ direction and diminish it by a factor α in the x direction. Flip the whole thing around $x = \frac{1}{2}$ and it will very closely resemble the lower part of figure 3.2 between λ_∞ and λ_1'.

In figure 3.7 a further magnification of the chaotic region around the period three window is shown. As can be seen, the period three orbit bifurcates into a period six orbit, etc. From the universality arguments this should be no surprise since the period three orbit may be formally reduced to a period one orbit of the function $f^3(x)$. Consequently, one expects, and indeed finds, that the approach to the relevant accumulation point is governed by the Feigenbaum constants δ and α.

One might then expect that the reversed bifurcations should also behave in a manner analogous to the reversed bifurcations of the main series. That would imply that the first bifurcation associated with this window should take place where the unstable period two hits the third and fourth iterates of the critical point. This takes place at about $\lambda = 0.9726$. As can be seen, in fact something happens at that point, but the spaces between the emerging branches are not empty. This phenomenon is called *crisis*. If we look closer at figure 3.7 we see that 'something' also happens at $\lambda \approx 0.969$. That is where the unstable period one hits the sixth iterate of the critical point. At about $\lambda = 0.9665$ the unstable period two hits the sixth and seventh iterates, and even further to the left period one hits the eighth iterate of the critical point and so on. All this has an accumulation point at approximately $\lambda = 0.9642$ where the crisis phenomenon ends.

3.10 Lyapunov exponents

Consider two orbits: one starts out at x_1 and the other an infinitesimal distance δx_1 away from x_1. Assume the orbits to be n time steps long, and n to be sufficiently small that the difference between the two orbits is still infinitesimal. Using the chain rule for differentiation, we find

$$\delta x_{n+1} = f'(x_n)f'(x_{n-1})\dots f'(x_1)\delta x_1. \qquad (3.47)$$

Define

$$\Lambda_n \equiv f'(x_n)f'(x_{n-1})\dots f'(x_1). \qquad (3.48)$$

Λ_n represents the relative change of the deviation of the two orbits from each other after n steps. It is customary to define

$$\lambda = \lim_{n\to\infty} \frac{1}{n}\ln|\Lambda_n| \qquad (3.49)$$

to be the *Lyapunov characteristic exponent*. (We shall use λ_{Ly} instead of just λ when there is any danger of confusion with the parameter of the logistic map.) It represents the coefficient of the average exponential growth per unit time of the relative distance between two infinitesimally separated orbits. The limit exists and is the same for *almost all orbits* or start points x_1.

We know that to the right of λ_∞ there is an infinity of stable periodic windows, i.e. periodic attractors. Outside the periodic windows there exist chaotic or strange attractors, but at any parameter value there exists only one attractor. On the other hand, at each parameter point with $\lambda > \lambda_\infty$ there is an infinity of unstable periodic orbits. Any such orbit will give an 'abnormal' result for the Lyapunov exponent in the sense that it will differ from the results for any orbit that starts on the attractor or in the attractor *basin of attraction*. A point belongs to the basin of attraction if an orbit starting at that point ends on the attractor, or has the attractor as a limiting set.

From the results in section 3.3 it follows that $\lambda_{\text{Ly}} = 0$ at a bifurcation point. Also at the superstable point there is one point, x, on the orbit where $f'(x) = 0$, implying that $\lambda_{\text{Ly}} = -\infty$. Thus for a stable orbit $\lambda_{\text{Ly}} < 0$ indicating that the phase space is *locally contracting*. For a chaotic orbit on the other hand $\lambda_{\text{Ly}} > 0$ indicating that the phase space is *locally expanding*. (This is a somewhat sloppy notion since nothing happens to phase space itself. What is meant is that a small section of phase space spanned by an infinity of nearby orbits is contracting.) The word 'locally' is crucial. We know that starting the orbit outside the interval $\langle 0, 1\rangle$ gives a diverging orbit even if there is a stable periodic attractor at that parameter point. On the other hand, the chaotic or strange attractor which a chaotic orbit

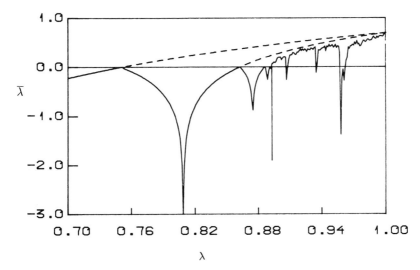

Figure 3.8 The Lyapunov characteristic exponent for the logistic map. The broken curves show the Lyapunov exponents for the unstable period one and period two orbits.

moves along will attract orbits that start outside the allowed intervals for that particular strange attractor.

We know from the numerical experiments of figures 3.2 and 3.5 that apparently whole intervals act as attracting sets, but once inside such a set, each point is locally repelling in such a way that two nearby orbits diverge exponentially from each other. This is called *sensitive dependence on initial conditions*, and is a fundamental property of all strange attractors and chaotic systems.

The Lyapunov exponent is very useful for scanning through a model without going through the often cumbersome task of looking at orbits at a large number of parameter points.

Figure 3.8 shows the Lyapunov exponent for the logistic map. The periodic orbits and windows show up as deep dips in the exponent, and a bifurcation occurs each time the exponent is zero. Since we know exactly the period one and two orbits it is a trivial matter to calculate their exponents. For the period one orbit we obtain

$$\lambda_{\mathrm{Ly}} = \ln|2 - 4\lambda| \qquad (3.50)$$

and for the period two orbit we find from equation (3.32):

$$\lambda_{\mathrm{Ly}} = [\ln|1 - (4\lambda + 1)(4\lambda - 3)|]/2. \qquad (3.51)$$

These exponents exist also outside the stability interval (broken curves in figure 3.8), but in that case there always exist other orbits with a lower Lyapunov exponent.

The Lyapunov exponent does not depend on the particular representation of the orbit, or equivalently the map. We demonstrate this by making the transformation

$$y = g(x) \tag{3.52}$$

where the new variable is y. We require that the transform has a continuous first derivative, and that a unique inverse exists for all x on the interval where the map is defined. In the new variable the map is

$$y_{n+1} = g(f(g^{-1}(y_n))). \tag{3.53}$$

To calculate the Lyapunov exponent in the new variable we need to calculate dy_n/dy_1. Because of the unique connection between x and y, we have

$$
\begin{aligned}
\frac{dy_n}{dy_1} &= \frac{dy_n}{dx_n} \frac{dx_n}{dx_{n-1}} \frac{dx_{n-1}}{dx_{n-2}} \cdots \frac{dx_2}{dx_1} \frac{dx_1}{dy_1} \\
&= g'(x_n)\Lambda_n/g'(x_1).
\end{aligned}
\tag{3.54}
$$

Since $g'(x_n)$ and $g'(x_1)$ are finite we have for the transformed Lyapunov exponent:

$$\lambda' = \lim_{n\to\infty} \frac{1}{n}[\ln|\Lambda_n| + \ln|g'(x_n)/g'(x_1)|] = \lambda \tag{3.55}$$

which is what we wanted to show. For a linear transformation the last term in (3.55) vanishes identically.

3.11 Statistical properties of chaotic orbits

Although chaotic orbits have sensitive dependence on initial conditions, their statistical properties are usually well defined and not dependent on initial conditions.

Consider a typical chaotic orbit at a parameter value to the right of the first reversed bifurcation. From geometrical considerations one easily sees that any starting point in the interval $\langle 0,1 \rangle$ will lead to a non-divergent orbit, thus there is an attractor present whose basin of attraction is simply the interval $\langle 0,1 \rangle$ excluding points that belong to unstable periodic orbits and their *pre-images* or *ancestors*. We also know from section 3.8 that the attractor must be confined to the interval $\langle 4\lambda^2(1-\lambda), \lambda \rangle$. By definition the orbit will never again visit a point once already visited, so to get to know the anatomy of the attractor somewhat better, we may divide the interval into small subintervals and count the number of hits in each subinterval, and thus obtain a distribution such as that in figure 3.9. It is seen from the figure that some regions are more popular than others.

At the particular parameter $\lambda = 1$ the shape of this distribution may be calculated also in a quite different way. Consider a point x and its pre-images x_1 and x_2 given by

$$x_{1,2} = (1 \pm \sqrt{1 - x/\lambda})/2. \tag{3.56}$$

From this one finds by differentiation the relation between the sizes of an infinitesimal region around x and corresponding regions around x_1 and x_2 :

$$|dx_{1,2}| = |dx|/(4\lambda\sqrt{1 - x/\lambda}). \tag{3.57}$$

Since any point in these two regions must be succeeded by a point in the corresponding region around x, one finds the following functional equation for the density of points $P(x) = dN(x)/dx$:

$$P(x) = (P((1 + \sqrt{1 - x/\lambda})/2) + P((1 - \sqrt{1 - x/\lambda})/2))/(4\sqrt{\lambda}\sqrt{\lambda - x}). \tag{3.58}$$

This functional equation is sometimes called the Perron–Frobenius equation. Remembering that λ is also the first iterate of the critical point, one sees from equation (3.58) that the distribution must be singular at the first iterate. Using equation (3.58) repeatedly one sees that the distribution must be singular at every iterate of the critical point. At $\lambda = 1$ the situation is particularly simple since the first iterate gives one, and all subsequent iterations give zero. Thus at $\lambda = 1$ there are singularities at $x = 0$ and at $x = 1$. The function

$$P(x) = 1/\sqrt{x(1 - x)} \tag{3.59}$$

is easily seen to satisfy equation (3.58). Also at the first band splitting bifurcation point things are simple since there are only three singular points in the distribution. The result of sampling from one chaotic orbit at this point is shown in figure 3.9. Figure 3.10 on the left shows the distribution of points from a single orbit right after the crisis has started.

Equation (3.58) is a simple means to gain insight into the nature of the shadowing of the bands of the chaotic region of figures 3.2, 3.5 and figure 3.7. Note that equation (3.58) is linear and thus can tell us nothing about the normalization.

3.12 Dimensions of attractors

It can be proved that the measure of the set of points constituting any— necessarily infinitely long—chaotic orbit is non-zero. This immediately implies that the fractal dimension of the chaotic attractor is one. It is also self-evident that the dimension of a periodic attractor is zero since

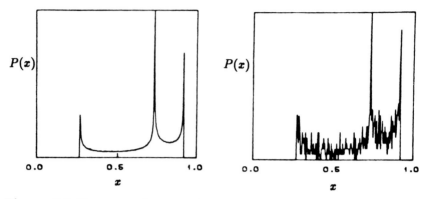

Figure 3.9 Histogram of 500 000 iterates of a single orbit approximating the density function $P(x)$ at $\lambda = 0.919\,643$ approximately at the first reversed bifurcation of the main series. The histogram to the left has been calculated using computer double precision for the iterations, while the orbit of the histogram to the right has been calculated using computer single precision.

the number of boxes needed to cover the points of the attractor will not grow when the length scale is reduced below a certain limit. One might think from this that the only results would be one and zero. However, this is not so. At an accumulation point the orbit is neither periodic nor truly chaotic, and numerical estimates, using the box counting algorithm described in chapter 2, give $D(\lambda = \lambda_\infty) = 0.538\ldots$. It is also conjectured that this is a universal number in the sense that this would be the dimension of the attractor at any period doubling cascade accumulation point almost independent of $f(x)$.

Since the attractor at the accumulation point can be constructed in a way very similar to the construction of the Cantor set in section 2.1 we can use this to make a crude estimate of the dimension of the attractor at λ_∞. Start by considering the interval $\langle f(\frac{1}{2}), f^2(\frac{1}{2})\rangle$. The whole attractor must evidently be inside this interval. Remove from this the interval between the third and fourth iterates of the critical point. The attractor must still be inside what remains. The process can be continued in a self-similar fashion *ad infinitum*. The self-similarity is not globally exact and the intervals that are removed are not removed from the middle of the parent interval, so taking b of equation (2.3) to be

$$b = \frac{f^3(\frac{1}{2}) - f^4(\frac{1}{2})}{f(\frac{1}{2}) - f^2(\frac{1}{2})} \tag{3.60}$$

must be a fairly crude approximation. The result is $D \approx 0.545$. If the self-similarity had been globally exact then the length of the two resulting intervals would have been $1/\alpha$ times that of the parent interval. This gives

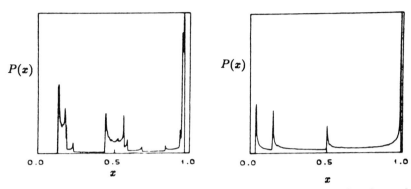

Figure 3.10 To the left the same as figure 3.9 at $\lambda = 0.9645$, just beyond the point where the phenomenon of crisis starts. To the right the same as figure 3.9 at $\lambda = 0.99$ just below the period four window.

$b = 1 - 2/\alpha$ and $D \approx 0.755$, which evidently is a very bad estimate.

3.13 Tangent bifurcations and intermittency

The equation to find period three orbits is $x = f^3(x)$. Having removed the two obvious solutions corresponding to the unstable fixed points $x = 0$ and $x = 1 - 1/(4\lambda)$ what remains is a sixth degree equation which can be solved exactly. The solution contains one period three orbit that is unstable in the whole interval where it exists, and another period three orbit that has a parameter interval where it is stable which starts at $\lambda = \lambda_c \equiv (1 + 2\sqrt{2})/4 \approx 0.957\,1068$. At that point the stable and the unstable orbits coincide. Exact results for the period three orbit are shown explicitly in appendix 1.

In figure 3.11 the function $f^3(x)$ at $\lambda = \lambda_c$ is displayed. The points $x = 0$ and $x = 1 - \frac{1}{4}\lambda$ are the points where the function crosses the line $x_{n+3} = x_n$. The other three solutions are all tangent points. For λ just below λ_c the only real solutions are the fixed points, while just above λ_c there are in addition six real solutions out of which it is easy to see from the geometry that three must be stable and thus belong to the same orbit. Since this stable orbit first occurred as a tangent solution these bifurcations are called *tangent bifurcations*, and this is the canonical way the periodic windows of the logistic map start out.

Define $\varepsilon \equiv (\lambda_c - \lambda)/\lambda_c$ and let ε be small and positive. Near the tangent points there will be very narrow channels as illustrated in figure 3.12, which is an enlargement of the central section of figure 3.11 for a non-zero value of ε. The orbit indicated in the figure belongs to the map $x \leftarrow f^3(x)$. This

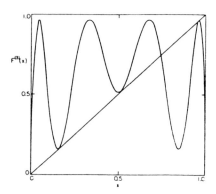

Figure 3.11 The function $f^3(x)$ at $\lambda = \lambda_c$.

is equivalent to recording only every third point of an orbit of the original
map. This process is often called *coarse-graining*. The orbit will spend a
long time nearly trapped in the channel, but once the orbit gets out of the
channel it will oscillate wildly till it again enters a channel which may be
any one out of the three. The quiet period spent in the channels is called the
laminar phase, while the occurrence of the wild oscillations is called *bursts*,
and the whole phenomenon of changes between a laminar phase and sudden
bursts is called *intermittency*. In physical systems intermittency seems to
be a common route to chaos.

For ε very small one may make a simple estimate of the time spent in
the laminar phase. An expansion of $f^3(x)$ in the relevant region gives the
following generic form:

$$x' = \varepsilon + x + ax^2 + \ldots. \tag{3.61}$$

The coefficient of the first-order term is one by definition. Since each step is
very small one may approximate $x' - x$ by dx and since the number of steps
through the channel is very large we may write $dn = 1$. These assumptions
transform equation (3.61) into an integrable differential equation:

$$dx/(\varepsilon + ax^2) = dn. \tag{3.62}$$

This gives
$$\tan^{-1}(x_1\sqrt{a/\varepsilon}) - \tan^{-1}(x_0\sqrt{a/\varepsilon}) = n\sqrt{a\varepsilon}. \tag{3.63}$$

Here x_1 is the positive end point of the channel and correspondingly x_0 is
the negative starting point. It is, of course, somewhat arbitrary where one
defines the channel to start and end, but since x_0 and x_1 have opposite
signs, the left-hand side of equation (3.63) will be finite and of order one

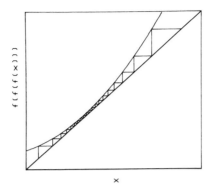

Figure 3.12 Enlargement of the central part of figure 3.11 for λ just below λ_c.

provided $|x_{1,0}| \gg \sqrt{\varepsilon/a}$. This means that the time spent in the channel is growing as $1/\sqrt{\varepsilon}$. A similar calculation shows that the Lyapunov exponent approaches zero like $\sqrt{\varepsilon}$.

3.14 Exact results at $\lambda = 1$

At $\lambda = 1$ one may compute all orbits exactly in terms of elementary functions. Since $x \in [0, 1]$ we may put

$$x = \sin^2 \theta \qquad (3.64)$$

which gives

$$\sin \theta_{n+1} = \pm \sin 2\theta_n. \qquad (3.65)$$

Any choice of sign will do, so we choose

$$\theta_{n+1} = 2\theta_n \pmod{2\pi} \qquad (3.66)$$

bearing in mind that the correspondence between x and θ is one to two. Defining $z \equiv \theta/2\pi$ we obtain

$$z_{n+1} = 2z_n \pmod 1. \qquad (3.67)$$

Suppose we write z in the binary representation then multiplication by two just means shifting everything one place to the left. This is known as a *Bernoulli shift*. The right-hand side of equation (3.67) is to be taken *modulo* 1 which means that one has to delete any integer part that might arise. This implies that the iterating function is not continuous. (See figure 3.13.)

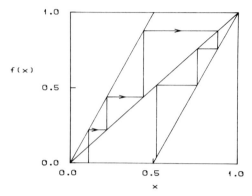

Figure 3.13 Graphical illustration of the iterative process using equation (3.67).

Let us look for periodic orbits. The simplest is obviously the or-
bit consisting of the point $z = 0.0000\ldots$. This is the same as $z = 0.1111\ldots \bmod 1 = 1 \bmod 1 = 0$. The next simplest orbit has period two and consists of the two points $z = 0.010\,101\ldots = \frac{1}{3}$ and
$z = 0.101\,010\ldots = \frac{2}{3}$.

Period three is somewhat more complicated. Taking the first point to
be $z = 0.001\,001\,001\ldots = \frac{1}{7}$ one gets an orbit consisting of the set $\frac{1}{7}$, $\frac{2}{7}$,
$\frac{4}{7}$. Clearly there is another orbit starting at $z = 0.011\,011\,011\ldots = \frac{3}{7}$. The
points of this orbit are $\frac{3}{7}$, $\frac{6}{7}$, $\frac{5}{7}$ ($\frac{5}{7}$ is $2 \times \frac{6}{7} \bmod 1$). In general a periodic
binary number of periodicity n is just a geometric series. Summing this
series one finds that the sum must have the form $p/(2^n - 1)$, and the
other members of the orbit are found by multiplication by two mod 1. In
principle there are $2^n - 2$ possible values of p, but not all corresponding z
values will generate true period n orbits. The situation can be illustrated
by looking at the $n = 4$ case. Suppressing the denominator 15 one has the
following four sequences:

$$
\begin{array}{llllll}
\text{A}: & 1 & 2 & 4 & 8 \\
\text{B}: & 3 & 6 & 12 & 9 \\
\text{C}: & 5 & 10 & 5 & 10 \\
\text{D}: & 7 & 14 & 13 & 11.
\end{array}
$$

Evidently C is only a period two orbit. From figure 3.4 we expect only two
period four orbits. The reason for this apparent contradiction is that in x
space the cases A and D coincide.

If the start point of an orbit has a binary representation which has
some arbitrary sequence of ones and zeros before the periodic sequences
start, it will be a case of an orbit being *trapped* or *captured* on an unstable
periodic orbit. All starting points which in the binary representation consist
of an initial sequence of numbers prior to a given periodic sequence will

eventually end up on the same orbit. Thus an orbit will not know where it came from: there is a *loss of memory*. The reason is that the map belongs to the class of maps that are *non-invertible (endomorphisms)*. Suppose z is an irrational number. Since there is no periodicity in z we may assume (i.e. *ergodicity*) that the orbit eventually will have visited equally many times all finite intervals in z of the same length. Thus the density of points of a chaotic orbit must be constant in z- and θ-space. With the notation from section 3.11 we get the density of points in x-space:

$$dN/dx = (dN/dz)/(dx/dz) \propto 1/(dx/d\theta). \tag{3.68}$$

Since

$$dx/d\theta = 2\sin\theta\cos\theta = 2\sqrt{x(1-x)} \tag{3.69}$$

we recover the distribution (3.59) in a very different way.

Combining equation (3.48) with equation (3.49) the Lyapunov exponent may be written as an infinite sum:

$$\lambda_{\text{Ly}} = \lim_{n\to\infty} \frac{1}{n} \sum_{i=1}^{n} \ln|f'(x_i)|. \tag{3.70}$$

Replacing the sum with an integral weighted with the density of points we obtain

$$\lambda_{\text{Ly}} = \int \frac{dN}{dx} \ln|f'(x)|dx \Big/ \int \frac{dN}{dx}dx. \tag{3.71}$$

Inserting $dN/dx = 1/\sqrt{x(1-x)}$ one finds $\lambda_{\text{Ly}} = \ln 2$. This is exactly the same as the exponent for the unstable period one and two at $\lambda = 1$ from equations (3.50) and (3.51). Also the once stable period three orbit gives the same number, so one might think that this is a general result. However, it is not since the period one orbit $x = 0$ gives $\lambda_{\text{Ly}} = 2\ln 2$.

The result for the chaotic attractor follows much more simply if we use the result from section 3.10, where it was shown that the Lyapunov exponent is independent of the representation. Using equation (3.67) we have $f'(x) = 2$ at all points of the orbit. This gives $\Lambda_n = 2^n$ and the previous result follows immediately.

3.15 Predicted power spectra. Critical exponents. Effect of noise

In this section a few important topics will be mentioned, but we shall refrain from going into any details.

In physics it is customary to measure *power spectra* instead of points in phase space. For our map this would mean measuring:

$$a_k = \lim_{N\to\infty} \frac{1}{N} \sum_{n=1}^{N} e^{(2\pi i k n)/N} x_n \quad k \in \{1, 2, \dots, N-1\}. \tag{3.72}$$

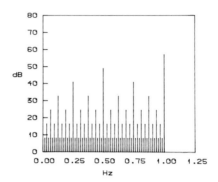

Figure 3.14 Expected power spectrum near λ_∞.

At $\lambda = \lambda_\infty$ the result is schematically illustrated in figure 3.14. The frequencies are given by the period doubling, *subharmonic*, sequence. In addition, the subharmonic amplitudes are related to each other by a *universal constant factor* $\mu = 6.57\ldots$ depending solely on α. On a logarithmic scale like that in figure 3.14 this transforms into a *constant difference* of $10\log_{10}\mu = 8.18\ldots$ dB which measures the amplitude reduction from one generation to the next. For a comparison between theory and experiments see figure 3.17.

In section 3.13 it was mentioned that near the intermittent transition to chaos the Lyapunov exponent behaves like $(\lambda_c - \lambda)^{1/2}$. Thus the *critical exponent* for this transition is $\frac{1}{2}$. For the period doubling route to chaos one finds that the Lyapunov exponent behaves like $(\lambda_\infty - \lambda)^t$ for λ immediately above λ_∞. The exponent $t = 0.449\,8069\ldots$ is a universal number.

In the presence of *external noise* the map may be changed into a *Langevin equation*:

$$x_{n+1} = f(x_n) + \sigma\xi_n. \tag{3.73}$$

Here ξ_n are random numbers following some specified distribution, and σ is the parameter that regulates the noise level. It is intuitively obvious that noise makes it impossible to detect bifurcations beyond a certain number. In order to see one more generation one must lower the noise level by the constant factor $k = 6.6190\ldots$ (compare to μ) or $10\log k = 8.21\ldots$ dB.

3.16 Experiments relevant to the logistic map

The experiments that most directly reproduce the results of the logistic map are those that are done on *nonlinear circuits* (figure 3.15). This is

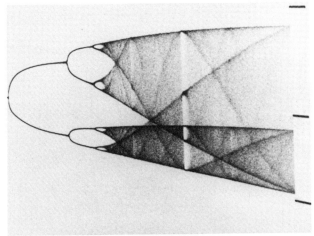

Figure 3.15 Results from a nonlinear circuit experiment. From Jeffries C D 1985 *Phys. Scr.* **59** 11. Reproduced with permission.

not much of a surprise since the circuits may be considered to be analogue computers solving a simple set of ordinary differential equations which one can solve numerically on digital computers and see the same bifurcation pattern as for the logistic map. (This bifurcation pattern can also be seen in the Lorenz model that we shall study in chapter 10.) Nevertheless, it is nice to find experimentally the universal numbers α, δ and μ within errors.

Totally non-trivial are the experiments on hydrodynamical instabilities in *Rayleigh–Benard convection cells*: a small container is heated carefully from below creating a temperature difference (the control parameter) between top and bottom. For small temperature differences there is no large-scale motion in the liquid, but at a certain critical temperature a convection flow sets in. The flow consists of a number of horizontal rolls, the number of which depends on the geometry of the container. When the temperature is increased further horizontal waves occur along the rolls (figure 3.16). This means that the temperature measured at one point will start to oscillate. Figure 3.17 shows the power spectrum of such oscillations. (Theoretical predictions are shown in figure 3.14.) Although the agreement with theory is good, the period doubling route to chaos is not the only one observed in this type of experiment.

3.17 Poincaré maps and return maps

Suppose that the system under consideration has a phase space of dimension higher than two, and that it evolves continuously in time in some non-trivial manner which makes its phase space trajectory stay confined

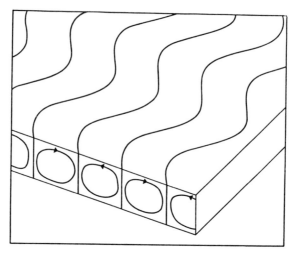

Figure 3.16 Sketch of the oscillatory instability in the Rayleigh–Benard experiment. From Clever R M and Busse F H 1974 *J. Fluid Mech.* **65** 625. Reproduced with permission.

within some finite volume.

To simplify the description of this evolution we may sacrifice some of the information and record only every time the trajectory intersects some subspace (in a three-dimensional case it could be the xy plane), thus creating a sequence of points where every point maps into the next in a space of reduced dimensionality. These types of maps are called Poincaré maps. If one is very lucky the points of the map may fall along some smooth curve indicating that a strong dimensional reduction has taken place.

Another way of reducing the complexity of the description is by using return maps. One may for instance consider only one variable, say x, and record it at some regular time interval, or every time the trajectory intersects some subspace, or every time it has some local extremum value, etc. The resulting sequence may then be used to plot x_{n+1} against x_n for all n. This often gives interesting information about the system. (In fact there exists a theorem saying that all necessary information may be extracted from an infinitely long time series of just one variable. See chapter 11.) Such a plot has to have at least one maximum point and one minimum point due to the fact that the trajectory is confined. If the points fall reasonably close to a curve with a smooth maximum, we may expect the system to have many of the features (like period doubling cascades) of the logistic map. This way of describing the system resembles what one has to do experimentally where it is impossible to observe but a few variables as compared to the enormous phase space dimensionality normally encountered in a real system.

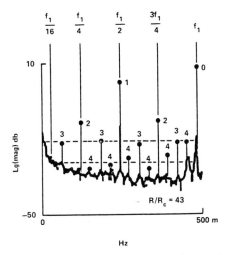

Figure 3.17 Power spectrum of a Rayleigh–Benard experiment near the transition to chaos. From Feigenbaum M 1980 *Los Alamos Science*, vol 1, 4. Figure partly redrawn from Libchaber A and Maurer J 1980 *J. Physique Coll.* **41** C3 51. Reproduced with permission.

3.18 Closing remarks on the logistic map

In a very handwaving way the most important argument for the *immediate* relevance of the logistic map to physics is essentially this: due to dissipation the phase space must be contracting (to a point if there is nothing driving the system). This contraction may be different in different directions and it may be expanding in some direction(s). In turn this will lead to the survival only of those directions that expand or at least do not contract. Although there are experimental examples of extremely drastic dimensional reductions it is nevertheless rare that it is so drastic that the logistic map becomes a very good model quantitatively. Consequently we are forced to look at higher dimensional systems, but the lessons learned from the logistic map will prove very valuable in these investigations.

4

THE CIRCLE MAP

The logistic map has a one-dimensional phase space and a one-dimensional parameter space. Its canonical routes to chaos are all period doubling. In this chapter we shall consider the circle map which has a one-dimensional phase space and a two-dimensional parameter space and its routes to chaos are far more complicated than in the case of the logistic map. The circle map is of physical relevance, for instance in describing the Josephson junction and other phenomena that involve recurring motion. Circle maps exibit periodic, quasiperiodic and chaotic motion in a complicated pattern. Many of the properties of the circle map are very closely related to the number of critical points, and not so much to the exact functional form of the map. A polynomial map of degree three and suitably chosen constants or parameters will have many of the same properties, but will not have quasiperiodic orbits.

The standard, dissipative circle map is given by the following common prescription:

$$\theta_{n+1} = \theta_n + 2\pi\Omega + K\cos(\theta_n + d) \pmod{2\pi}. \tag{4.1}$$

Here Ω, K and d are parameters. Often the $\sin(x)$ function is used instead of $\cos(x)$, but obviously the two are equivalent, only requiring d to be fixed to some particular and irrelevant value. Without loss of generality the parameter d may be dropped by making a translation in θ. The map equation (4.1) may be looked upon as a mapping of points on a circle onto the circle, hence the name circle map. The circle map is intimately related to *Hopf bifurcations* in higher dimensional maps when each of the points of a periodic attractor bifurcates into topological circles.

For many purposes it is more convenient to let the map be on the unit interval, so we write instead of equation (4.1)

$$x_{n+1} = f(x_n) \equiv x_n + \Omega + K\cos(2\pi x_n)/2\pi \pmod{1}. \tag{4.2}$$

4.1 The fixed points

The fixed points are determined by putting $x_{n+1} = x_n = x$, yielding

$$N = \Omega + K\cos(2\pi x)/2\pi \tag{4.3}$$

where N is an integer chosen so that x falls on the unit interval. Depending on the value of K and Ω there may be one, two or more solutions which may be stable or unstable. For K values that allow $N = 0$ one has

$$x = \cos^{-1}(-2\pi\Omega/K)/2\pi. \qquad (4.4)$$

Differentiating the right-hand side of equation (4.2) with respect to x_n gives the stability condition

$$-1 < 1 - K\sin(2\pi x) < 1. \qquad (4.5)$$

The first of these inequalities gives the period doubling transition, while the last inequality locates a transition where the period in principle becomes infinitely long. This transition is a tangent bifurcation, but is not neccesarily a transition to chaos. For small K it is a transition to quasiperiodicity. We shall return to this point in the last section of this chapter.

Using equation (4.4) we find the period doubling bifurcation along the hyperbola
$$K = \sqrt{4 + (2\pi\Omega)^2}. \qquad (4.6)$$
Thus the tangent bifurcation occurs along

$$K = \pm 2\pi\Omega. \qquad (4.7)$$

The lower part of the stable region looks like a wedge and equation (4.7) gives the wedge boundaries. (See plate 1 and figures 4.1 and 4.5.) Notice that although the above calculations show where the fixed point is stable in parameter space, they do not show that there may not be other stable orbits at the same parameter values. In other words, the basin of attraction of the fixed point need not be the whole phase space.

4.2 Circle maps near $K = 0$. Arnol'd tongues

Investigating the positions of other periodic orbits has to be done by approximate methods, but going back to equation (4.2) for the case $K = 0$ gives some important insight.

Consider the map for $\Omega = p/k$ wher p and k are relative primes. The map then reads
$$x_{n+1} = x_n + p/k \pmod 1. \qquad (4.8)$$
Starting at x_0 and iterating k times gives

$$x_k = x_0 + p \pmod 1 = x_0. \qquad (4.9)$$

Evidently all orbits at this point in parameter space are period k orbits independent of starting point in phase space. There may be many different

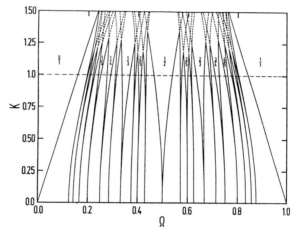

Figure 4.1 A few dominant Arnol'd tongues starting at $\Omega = 0$. When all tongues are included, the resonance overlap starts at the critical value $K = 1$. From Jensen M H *et al* 1984 *Phys. Rev.* A **30** 1960. Reproduced with permission.

choices of p for any given k, and period k orbits may be found at all rational values of Ω with denominator k. They are distinguishable by the order in which the points on the circle are visited. The ratio p/k is called the *winding number* of the orbit. More generally one may define a winding number for any K and Ω for any infinitely long orbit by

$$W = \lim_{n \to \infty} (x_n - x_0)/n. \tag{4.10}$$

In this definition x_n is computed without taking (mod 1).

Equation (4.10) shows that any fixed point with $N = 0$ will have the sensible winding number zero. We have seeen that the fixed point is stable in a region of finite width as soon as $K > 0$. Fixing K to some value less than one we may do numerical experiments and see that wedges of constant rotation numbers and finite widths are emanating from every rational Ω on the $K = 0$ axis. (See figure 4.1.) These wedges are more commonly referred to as *Arnol'd tongues*. For any $K \in \langle 0, 1 \rangle$ there exist finite intervals in $\Omega \in \langle 0, 1 \rangle$ representing all possible rational winding numbers less than one. Furthermore, their widths are growing monotonically with K, and still they do not overlap till $K > 1$. At first sight it might seem strange that the sum of infinitely many finite pieces add together to something finite. However, this is no more peculiar than the convergence of a simple geometric series. All that is needed is for the width to fall off sufficiently rapidly as a function of k for a given K. Just for the sake of argument, suppose the widths grow with K like K^k. There are no more than k different p values, so the total

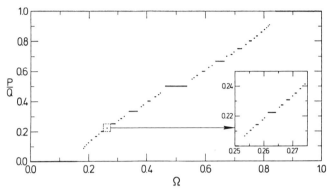

Figure 4.2 The Devil's staircase at $K = 1$. Under magnification the staircase nearly reproduces itself. From Jensen M H *et al* 1984 *Phys. Rev.* A **30** 1960. Reproduced with permission.

length of the the stable orbit intervals is certainly less than

$$l = \sum_{k=1}^{\infty} kK^k. \tag{4.11}$$

This sum is simple to calculate and gives a finite result for $K < 1$. However, in reality the correct sum is finite also at $K = 1$ and so the widths of the Arnol'd tongues grow less rapidly with K than indicated above, at least near $K = 1$.

We can now make an interesting numerical experiment by calculating and plotting the winding number against Ω for some fixed K. (See figure 4.2.) The result is often called the *Devil's staircase* for reasons left to be figured out by the reader. As can be seen, by suitably stretching and scaling one small piece of the staircase, one may obtain a figure that is similar to the whole staircase. For $K < 1$ the total length of the regions of stable periodic orbits is less than one and therefore the staircase is said to be incomplete. At $K = 1$ there is numerical evidence that the total length is one and the staircase is complete.

It is useful to think of the rational winding numbers as the ratios of two competing frequencies, one for the system, and one for some external driving force. The phenomenon that the winding number locks itself onto a fixed rational number for a finite interval in parameter space is called *phase locking, mode locking, frequency locking, resonance* or *entrainment* and is a very important notion in many dynamical systems. It was first described by Huygens, who observed that two clocks that did not move with exactly the same speed kept at a distance from one another, in fact locked to the same freqency when put back to back, due to the tiny coupling to each other.

Suppose that K is just infinitesimally greater than zero and consider the Lyapunov exponent given by equation (3.49). Since K is infinitesimal one has

$$\sum_i \ln(f'(x_i)) \approx \sum_i -K \sin(2\pi x_i). \qquad (4.12)$$

For periodic orbits the sum extends only over a finite number of points and for those orbits the Lyapunov exponent is proportional to K. For the stable periodic orbits the Lyapunov exponent must be negative. If Ω is chosen so that the orbit has an irrational winding number one may asume that the vicinity of all points is visited equally often. The sum in equation (4.12) may then be replaced by an integral which obviously stays finite and small, so that the corresponding Lyapunov exponent is zero. Orbits that are not periodic and have the Lyapunov exponent equal to zero are said to be *quasiperiodic*. For small K most values of Ω lead to quasiperiodic orbits. Suppose W is an irrational number. It may then be approximated by two rationals, one just below, and the other one just above W. By making better and better approximations, one may localize orbits of winding number W with any desired precision forming a continuous line in the (Ω, K) plane starting at $(\Omega = W, K = 0)$. Notice that for points on this line $\Omega = W$ is true only at $K = 0$. For $K < 1$ these orbits all have Lyapunov exponent zero, and are said to be *quasiperiodic*. (See figure 4.7.) For small K most orbits are quasiperiodic. This should be compared with the fact that the measure of all rationals on the interval is zero.

4.3 The critical value $K = 1$

Going to the critical value $K = 1$ (See figure 4.8 and compare with figure 4.7) there is numerical evidence that the measure of the resonance regions is one. This does not mean that the quasiperiodic orbits are totally absent, but they are found on a set of zero measure. The way the numerical evidence for this comes about is interesting. Choosing a particular measuring stick of length r and covering the whole interval with sticks (one-dimensional boxes) of this length, one may count the number $N(r)$ of sticks that cover more than one resonance. As r decreases, $N(r)$ increases according to a power law, as can be seen in figure 4.3. In accordance with the prescription of chapter 2, this allows one to calculate a fractal dimension. The result is $d \approx 0.870$. Since this is less than one it follows that the topological dimension and the measure of the set is zero, and hence what remains has measure one.

For K less than the critical value, the winding number is unique, independent of the initial point of the orbit. Above the critical value there may be two different results depending on initial conditions due to the possibility of coexistence of two attractors at the same parameter value.

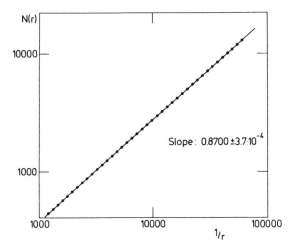

Figure 4.3 Number of 'holes' in the resonance region at the critical value $K = 1$ as a function of the length of the measuring stick. The resulting dimension is $d \approx 0.870$. From Jensen M H *et al* 1984 *Phys. Rev.* A **30** 1960. Reproduced with permission.

4.4 Period two, bimodality, superstability and swallowtails

For $K < 1$ things were simple and qualitatively easy to understand. For $K > 1$ the dynamics become much more complicated. Consider

$$f'(x) = 1 - K \sin(2\pi x). \tag{4.13}$$

It is obvious that $f'(x)$ is positive for all x and $f(x)$ is monotonically increasing as long as $K < 1$. Therefore the map has no critical points for $K < 1$. $K = 1$ is said to be a critical line in the parameter plane. For $K' > 1$ it is easy to see geometrically that there are two critical points. Their positions are found by solving

$$1 - K \sin(2\pi x) = 0. \tag{4.14}$$

When a map has two critical points, it is said to be *bimodal*. The existence of more than one critical point has some important consequences, one being that it may be possible to have more than one attractor at a given point in parameter space. If $f(x)$ is a rational function there exists a theorem by Fatou saying that *the number of attractive orbits is at most equal to the number of critical points*. Although our $f(x)$ is not a rational function it is still true that the number of coexisting, stable periodic orbits are never more than two. (See figure 4.4 and plate 1.)

Just as in the case of the logistic map, it is very useful to study the iterations of the critical points. However, since the parameter space is two

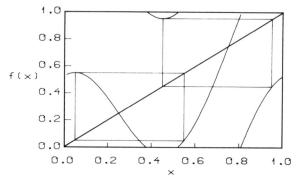

Figure 4.4 Coexisting period two orbits at $K = 3.3, \Omega = 0.0$. The orbits are very nearly superstable.

dimensional, the results will be surfaces in the three-dimensional space of the two parameters and the one-dimensional phase space. These are not particularly useful to picture. It is more instructive to study the intersections of these surfaces with the surface defined by $f'(x) = 0$. This will be in the (Ω, K) plane where the orbits are superstable. (See figure 4.5.) The order of the iteration will determine the periodicity of the orbits.

In the case of the period one orbit the position of the superstable orbits in the parameter plane is easily calculable in just the same way as we calculated the position of the period doubling bifurcation. The result is

$$K = \sqrt{1 + (2\pi\Omega)^2}. \tag{4.15}$$

To get some further insight it is very useful to calculate the Lyapunov exponent at very many points in parameter space and colour the corresponding points in a graph according to the results. Such a picture is often called a *phase diagram*, and may be produced in other ways than using Lyapunov exponents. A good phase diagram reduces to a minimum the tiresome process of looking at phase space orbits. The result of this kind of numerical experiment is shown in plate 1. One can easily see the large region where the fixed point is stable, and the hyperbola where the period doubling bifurcation takes place. However, the next period doubling is not so trivial because a 'swallowtail' structure emerges. This is because there are two different period two orbits (see figure 4.7 and figure 4.8), each with its own basin of attraction, and further period doublings seem to involve the same type of 'swallowtail' structures. It is also quite obvious from the same experiment that the resonances overlap, meaning that the result, i.e. the attractor, depends on the initial point of the orbit and the basin of attraction may become complicated. However, since there are only two critical points, no more than two resonances or Arnol'd tongues may overlap at the same parameter point. As K increases chaotic orbits

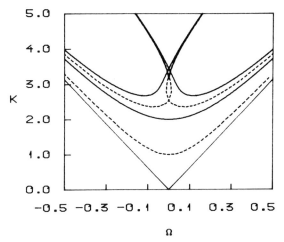

Figure 4.5 The bifurcation patterns of the period one winding number zero orbit and the positions in parameter space of the resulting period two orbits, both of winding number zero. Full curves represent bifurcations, both period doubling bifurcations and tangent bifurcations. The broken curves are curves where there exist at least one superstable orbit. The upper broken curves are curves where the second iterate of the critical points is zero, thus these lines show where there exist superstable period two orbits. These curves cross and form the centre of the characteristic swallowtail structure. Period two structures of winding number $\frac{1}{2}$ starting as Arnol'd tongues at $K = 0$ are not shown.

become more and more dominant in a fashion that can be read off from the Lyapunov exponent phase diagram in plate 1.

Because paper drawings are only two dimensional, it is often neccesary to sacrifice or fix one or more interesting variables. One may then get graphs that are very much a result of how one cuts in the parameter plane. An example is shown in figure 4.6 where K is kept fixed. Then Ω is varied and the phase space plotted vertically. The result is some bubbles—the consequence only of cutting straight through structures that are curved in the parameter plane.

4.5 Where can there be chaos?

There is an important theorem by Arnol'd saying that as long as $f(x)$ is smooth and invertible, no chaos can occur. In our case $f(x)$ is monotonically increasing as long as $0 < K < 1$ (apart from the jump due to the (mod 1) rule) and therefore the map is invertible, and hence there can be no chaos for $K < 1$. We found earlier that the fixed point was stable in a region that extended far above the critical parameter line $K = 1$, so the

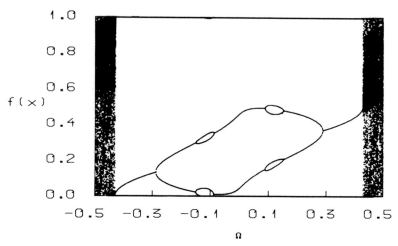

Figure 4.6 Diagram showing the result of plotting 200 consecutive x values against Ω at fixed $K = 2.7$. Compare with figure 4.9.

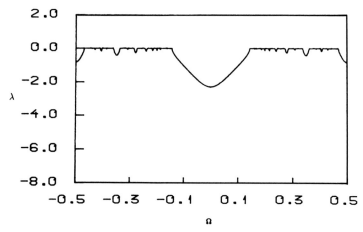

Figure 4.7 The Lyapunov exponent at $K = 0.9$ as a function of Ω. See plate 1.

the requirement $K > 1$ for the attractor to be chaotic is only necessary, but not sufficient. The stable region of the fixed point was limited from above by a hyperbola given by equation (4.7) with its lowest point at $K = 2$. Resonances with higher periodicity have Arnol'd tongues that are limited from above by hyperbola-like curves with their minima coming closer to the critical value $K = 1$ as the periodicity increases. For irrational winding numbers there is a transition from quasiperiodicity to chaos at $K = 1$, characterized by the Lyapunov exponent that changes from zero to some finite

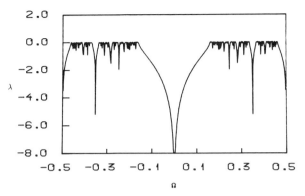

Figure 4.8 The Lyapunov exponent at $K = 1.0$ as a function of Ω. See plate 1.

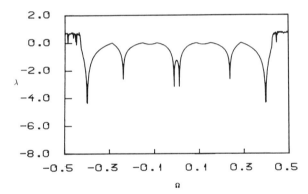

Figure 4.9 The Lyapunov exponent at $K = 2.7$ as a function of Ω. See plate 1.

value as in a second-order phase transition. To do numerical experiments on how quasiperiodic orbits make the transition to chaos, it is desirable to choose irrationals that do not have dominant periodic orbits as near neighbours. A pet irrational has been the golden mean $w = (\sqrt{5} - 1)/2$. It has the continued fraction expansion

$$w = 1/\{1 + 1/[1 + 1/(1 + \cdots)]\}. \tag{4.16}$$

At the golden mean winding number the Lyapunov exponent is found to change according to a power law $\lambda = (K - 1)^\gamma$ where $\gamma \approx 0.948$. It would have been nice if this exponent had been universal for all irrational winding numbers, but this is not so.

Following the parameter space along one of the swallowtails, one attractor stays topologically unchanged, while the other undergoes successive bifurcations into chaos. Thus chaotic and resonant attractors may coexist.

Returning to the question in the first section, what kind of transition takes place along the lines given by equation (4.7)? We can now say that for $K < 1$ it is a transition to quasiperiodicity, while for $K > 1$ there are two possibilities: either it is a transition to chaos, or it is a transition into another periodic orbit—the periodicity depends on Ω. The last type of transition is closely related to the phenomenon of overlapping resonances (see figure 4.1). Numerical support for this may be found by comparing the Lyapunov exponents of figures 4.7, 4.8 and 4.9.

5

HIGHER DIMENSIONAL MAPS

However useful the one-dimensional logistic map and the circle map have proved to be, it is necessary also to study higher dimensional maps. In particular, there have been many studies of two-dimensional maps. We shall only consider in detail the Hénon map, the linearly coupled logistic map and the conservative twist map. We shall also have a brief look at the complex logistic map. However, in this chapter we shall look into some of the more general aspects of higher dimensional maps.

5.1 Linear maps in higher dimensions

As in the one-dimensional case we shall start by considering the linear map in n dimensions because it describes approximately the behaviour of nonlinear maps near fixed points.

The state vector at time t is called $\boldsymbol{x}_t$, and its ith component is x_t^i. We shall assume that the space is real so that all components of the phase space vector are real at all times. Initially we consider only the case when $\boldsymbol{x}_t$ depends solely on $\boldsymbol{x}_{t-1}$. By a suitable definition of the vector space, one can make sure that $\boldsymbol{x}_t$ formally never depends on times earlier than $(t-1)$. However, it is often more practical to let there be dependence also on earlier times.

The general linear map is then given by

$$x_{t+1}^i = \sum_{j=1}^{n} A_{ij} x_t^j + b^i. \tag{5.1}$$

Here A is a constant matrix and $\boldsymbol{b}$ is a constant vector. Provided the matrix $1 - A$ is non-singular, $\boldsymbol{b}$ may be translated away. We shall assume this to be possible and hence shall consider equation (5.1) without the constant vector. This means that the null vector or the origin is a fixed point of the map.

We shall assume that A is real and non-singular. The eigenvalue equation will then contain only real coefficients. This means that the eigenvalues may be of two types, either real or complex. However, any complex eigenvalue is accompanied by another eigenvalue which is its complex conjugate.

This is seen simply by complex conjugating the eigenvalue equation. Any eigenvalue, λ, may then be written as an exponential

$$\lambda = e^{\rho + i\omega} = e^{\rho}(\cos\omega + i\sin\omega). \tag{5.2}$$

For real λ the corresponding eigenvector, e_λ, has only real components. Suppose the initial vector x_0 is proportional to the eigenvector e_λ. Clearly this is possible only for real eigenvectors. The time dependence is given by using equation (5.1) repeatedly, and the result is

$$x_t = \lambda^t x_0. \tag{5.3}$$

Depending on whether $|\lambda|$ is smaller or greater than one, the motion is either exponentially growing or contracting in the one-dimensional space spanned by the eigenvector.

If λ is complex, the corresponding eigenvector has complex components, and it is not possible to use it in a decomposition of the initial vector without also using its complex conjugate companion. From the complex eigenvector and its complex conjugate one may span a real, two-dimensional space using the real and linearly independent vectors

$$e_R = \tfrac{1}{2}(e_\lambda + e_\lambda^*)$$
$$e_I = \tfrac{i}{2}(e_\lambda - e_\lambda^*). \tag{5.4}$$

Consider now an initial vector in the plane spanned by these two vectors. The time evolution must then be given by

$$x_t^i \propto e^{\rho t} \cos(\omega t + \phi_i). \tag{5.5}$$

Here the phases ϕ_i are fixed by the initial conditions. This shows that complex eigenvalues imply a spiralling motion in a two-dimensional plane. The spiralling is outwards if $\rho > 0$ and inwards if $\rho < 0$. If $\rho = 0$ the motion is confined to an ellipse in the plane.

5.1.1 The manifolds of the fixed point

Because of the linearity the motion may be decomposed into exponentially behaving components associated with real eigenvalues, and exponential spirals or ellipses in two-dimensional planes spanned by the vectors associated with complex eigenvalues. The subspace spanned by those eigenvectors that have positive, zero or negative ρ is said to span respectively *the unstable manifold, the centre manifold* or *the stable manifold* of the fixed point. A fixed point that has only a stable manifold is said to be stable. A fixed point that has a centre manifold and no unstable manifold is said to be *elliptic*. A fixed point that has $\rho \neq 0$ for all eigenvalues is said to be *hyperbolic*. A fixed point that has some $\rho > 0$ and some $\rho < 0$ is said to be a *saddle point*. (The concept hyperbolic point is sometimes used as a synonym of a saddle point.) When $\omega_{1,2} = 0$ in two dimensions, the fixed points are called a *stable node* if $\rho_{1,2} < 0$, a *saddle node* if $\rho_1 < 0 < \rho_2$ and an *unstable node* if $\rho_{1,2} > 0$.

Consider as an example a three-dimensional case where there is one real eigenvalue with $\rho > 0$ and its corresponding eigenvector is along the z axis. Furthermore there is a complex conjugated pair of eigenvalues that spans the xy plane, and the corresponding ρ is less than zero. Suppose the initial point is far away from the z axis and just slightly above the xy plane. Initially the motion will be dominated by the spiralling motion towards the z axis, but sooner or later the orbit will be ejected along the z axis away from the origin. This type of motion is seen in many interesting models where the nonlinearity takes the system away from the z axis and back again to the vicinity of the xy plane.

5.1.2 Dependence on more than one time step

Let the phase space be one dimensional and the time evolution given by

$$x_{t+1} = \sum_{i=1}^{n} a_i x_{t+1-i}. \tag{5.6}$$

The n constants a_i and the n initial conditions determine the motion uniquely. Let us make the ansatz that $x_t = k^t$, and k is a constant to be determined. Inserting into equation (5.6) one obtains the nth order equation

$$k^n - \sum_{i=1}^{n} a_i k^{n-i} = 0. \tag{5.7}$$

This equation has n different solutions for k, so the general expression for the time dependence of x_t is

$$x_t = \sum_{i=1}^{n} c_i k_i^t. \tag{5.8}$$

Here we have labelled the different values of k with an index i so that $|k_i| \geq |k_{i+1}|$ and introduced n constants c_i to be determined by the initial conditions. If we consider very long times, the component generated by k_1 will dominate, and the other components will be comparably negligible.

The famous Fibonacci series is a map of the type equation (5.6) given by

$$F_{t+1} = F_t + F_{t-1}. \tag{5.9}$$

The initial conditions are $F_0 = 0$ and $F_1 = 1$. Solving equation (5.7) for this problem, one finds $k_{1,2} = \frac{1}{2}(1 \pm \sqrt{5})$. Using equation (5.8) for $t = 0$ and $t = 1$, one obtains

$$F_t = \frac{1}{\sqrt{5}}(k_1^t - k_2^t). \qquad (5.10)$$

In particular it follows that

$$\lim_{t \to \infty} F_{t+1}/F_t = k_1. \qquad (5.11)$$

This is the golden mean. (Actually it is the inverse of what we called the golden mean in chapter 4 about the circle map.)

The types of motion we obtain in the cases equation (5.1) and equation (5.6) are clearly very similar, and indeed, it is easy to show that equation (5.6) is just a special case of equation (5.1). This may be done by introducing a new, n-dimensional phase space. The new state vector has components $y_t^i = x_{t-i+1}$ and the map equivalent to equation (5.6) is given by

$$
\begin{aligned}
y_{t+1}^1 &= \sum_{i=1}^{n} a_i y_t^i \\
y_{t+1}^i &= x_{t-i+2} = y_t^{i-1} \quad i = 2, \ldots, n.
\end{aligned}
\qquad (5.12)
$$

This is just a special case of equation (5.1).

One problem with all linear maps is that they describe motion that either ends up on the fixed point, or 'explodes', or performs some trivial oscillations. Therefore these maps are not of much practical use in the making of *global* models, but they are important to gain insight into the various possible modes of motion near fixed points of nonlinear models. However, if one adds to equation (5.6) 'noise' in the form of a stochastic term, the map becomes much more intersting and useful. This is called the *autoregressive model of order* n, AR(n), and is of much use in time series analysis. It will be discused in chapter 11.

5.2 Manifolds. Homoclinic and heteroclinic points

Let the N-dimensional nonlinear map be given by

$$x_{t+1}^i = F^i(x_t^1, x_t^2, \ldots, x_t^N) \quad i = 1, 2, \ldots, N \qquad (5.13)$$

which we shall write for short as

$$\boldsymbol{x}_{t+1} = \boldsymbol{F}(\boldsymbol{x}_t). \qquad (5.14)$$

The *Jacobian matrix*, j, of the map is given by

$$j_{ik} = \frac{\partial F_i(\boldsymbol{x})}{\partial x_k}. \tag{5.15}$$

If the Jacobian (determinant) is one at all points in phase space the map is said to be *conservative*, otherwise it is *dissipative*.

Consider one fixed point $\boldsymbol{x} = \boldsymbol{F}(\boldsymbol{x})$ of the map and linearize around this fixed point. Suppose the Jacobian matrix of the fixed point has a real eigenvalue of absolute magnitude greater than one. Let us choose a point in phase space at a relative position $\delta\alpha e_i$ from the fixed point, and e_i is the eigenvector of the Jacobian matrix corresponding to the eigenvalue Λ_i and $\delta\alpha$ is an infinitesimal constant. Iterating this point once brings us to a point $\Lambda_i\delta\alpha e_i$ also along the same direction. Imagine that an infinite sequence of iterations of the full nonlinearized map is made for every point on the straight interval between $\delta\alpha e_i$ and $\Lambda_i\delta\alpha e_i$. For $\Lambda_i > 1$ the resulting connected curve is normally infinitely long. If that is the case the curve must be infinitely many times folded. In conservative maps this type of subspace of the unstable manifold is often referred to as a *separatrix*. If the map is invertible, the separatrix or unstable manifold does not intersect itself.

If the eigenvalue $\Lambda_i < 1$, one might attempt a similar procedure as the one described above. However, in that case all orbits will approach the fixed point, and the resulting curve will only be infinitesimally long and end at the fixed point. If the map is invertible the stable manifold may be found by just mapping backwards in time. If the map is not invertible more sophisticated methods have to be used. All pre-images of points on the line segment previously defined belong to the stable manifold.

The full unstable manifold may now be thought of as the space generated by an infinity of iterations of all points within some infinitesimal volume element spanned by all eigenvectors corresponding to eigenvectors of the Jacobian matrix of absolute value greater than one. Correspondingly, the full stable manifold is the set of all preimages to points within some infinitesimally big volume element around the fixed point, and generated by all eigenvectors corresponding to eigenvalues of the Jacobian matrix of absolute value less than one, i.e. $\rho < 0$.

In the linear case we could define a centre manifold corresponding to eigenvalues with $\rho = 0$. In that case any orbit starting on the centre manifold will either map onto an ellipse or onto a finite number of points on an ellipse. The last case will arise if ω is a rational multiple of 2π. In the nonlinear case, orbits starting on the centre manifold will iterate into closed curves (topological circles) or to point sets. This phenomenon is very important for conservative maps.

To each of the unstable, stable or centre manifolds there exist corresponding linear *tangent spaces* generated by the relevant eigenvectors.

Normally any of the above defined manifolds are very complicated geometrical objects with an infinity of bends and folds. The stable and unstable manifolds may intersect at points called *homoclinic points*. If the intersecting stable and unstable manifolds belong to different fixed points the intersection points are called *heteroclinic points*.

Since a point on the stable or on the unstable manifold is mapped into a point on the same manifold it follows that a homoclinic point is mapped into another homoclinic point, and that there is an infinity of such points.

Let the map be two dimensional and consider a closed loop, and use the map once for very point on the loop to map the loop onto another curve which for continuous maps necessarily must be another closed loop. The area of the latter loop is given by

$$\oint \mathrm{d}y_{n+1}\, \mathrm{d}x_{n+1} = \oint |j|\, \mathrm{d}y_n\, \mathrm{d}x_n. \tag{5.16}$$

In the particular case that the Jacobian is a constant there is a constant ratio between the areas of successive loops. Of particular interest are the loops formed by the sections of the stable and unstable manifolds joining two homoclinic points. Consider for instance a conservative map for which $|j| = 1$, meaning that consecutive loops have the same area. Near a hyperbolic fixed point the distance between homoclinic points along the stable manifold decreases, and thus the shape of the loops must get more and more prolonged in the direction of the unstable manifold, and eventually must get folded into extremely complicated mazes.

For dissipative maps we must have $|j| < 1$, at least in some average sense for points along an orbit, and consequently consecutive loops will have smaller and smaller areas, but for a saddle point the loops still become infinitely prolonged.

5.3 Lyapunov exponents in higher dimensional maps

In section 3.10 it was demonstrated that the Lyapunov exponent is a very useful means to explore the general structures occurring in phase space as a function of the parameter of the logistic map. This is no less true in higher dimensional maps, but in this case the number of exponents is equal to the phase space dimension.

Consider the map $x_0 \rightarrow x_T$ where T is some large number. Using the chain rule and Einstein's summation convention (sum over repeated indices), we find the Jacobian matrix for that map

$$J_{ij}(x_0 \rightarrow x_T) = j_{il}(x_{T-1} \rightarrow x_T)j_{lm}(x_{T-2} \rightarrow x_{T-1}) \cdots j_{kj}(x_0 \rightarrow x_1) \tag{5.17}$$

or written in a different notation

$$J_{ij} = \frac{\partial x_T^i}{\partial x_0^j} = \frac{\partial x_T^i}{\partial x_{T-1}^m} \frac{\partial x_{T-1}^m}{\partial x_{T-2}^l} \cdots \frac{\partial x_1^k}{\partial x_0^j}. \tag{5.18}$$

Starting the orbit at a nearby point $x_0' = x_0 + \delta x$ one finds after T iterations $x_T' = x_T + \delta x_T$. Dropping indices one finds to first order in δx_0:

$$\delta x_T = J \, \delta x_0. \tag{5.19}$$

Unless all eigenvalues of J have an absolute value less than one the orbit must be linearly unstable. Suppose we diagonalize J and call its eigenvalues Λ_i $i = 1, \dots, n$ and arrange it so that $|\Lambda_k| \geq |\Lambda_{k+1}|$ for all k. Then the kth Lyapunov exponent is defined to be

$$\lambda_k = \lim_{T \to \infty} \frac{1}{T} \ln |\Lambda_k| \tag{5.20}$$

provided the limit exists. For all points x_0 within one basin of attraction this limit is *independent* of x_0.

Assume T to be very large but finite. Then

$$|\Lambda_k| \sim e^{\lambda_k T} \tag{5.21}$$

and consequently

$$|\Lambda_k/\Lambda_{k+1}| \sim e^{(\lambda_k - \lambda_{k+1})T}. \tag{5.22}$$

Clearly this ratio goes to infinity with T provided $\lambda_k \neq \lambda_{k+1}$. (Degenerate Lyapunov exponents (LEs) are quite normal.) Thus straightforward diagonalization algorithms on finite precision computers can normally not be used in the general case since it will imply calculating differences between almost equal numbers. Suppose that J has already been diagonalized somehow, then the trace is approximately equal to $e^{\lambda_1 T}$ since the other eigenvalues make an insignificant contribution to the sum. Since the trace is invariant we may find the largest eigenvalue by just taking the trace. To find the other Lyapunov exponents in general requires special techniques which we shall not describe here. For maps of dimension two and three an algorithm to calculate the Lyapunov exponents is given in appendix 2. Here we note that

$$\prod_{i=1}^{N} \Lambda_i = \text{Det } J. \tag{5.23}$$

In many interesting cases $\text{Det } j = B$ where B is a constant independent of time and position in phase space. This implies that $\text{Det } J = B^T$. In such cases we get

$$\sum_{i=1}^{N} \lambda_i = \ln |B|. \tag{5.24}$$

In the particular case that the phase space volume does not change during the iteration $|B| = 1$. So for *volume (or area) preserving maps*, i.e. conservative maps

$$\sum_{i=1}^{N} \lambda_i = 0. \tag{5.25}$$

Another consequence of equation (5.24) is the relation

$$\lambda_N \leq \frac{\ln |B|}{N}. \tag{5.26}$$

The geometrical meaning of the Lyapunov exponents is that for positive exponents there exist directions, rotating along the trajectory, in which the motion on the average is unstable or divergent in the sense that nearby trajectories in these directions will diverge from the original orbit. Although the orbit is unstable, its stable directions provide sufficient volume contraction that the orbit is confined to some bounded region in phase space. In such a case the orbit is said to be attracted to a *strange attractor*. When $\lambda_1 = 0$ the orbit is *quasiperiodic* and the motion takes place on a 'torus' in phase space. For instance in two dimensions this is the motion on one or more closed curves topologically equivalent to circles. When all Lyapunov exponents are negative the orbit ends on a stable fixed point or a stable periodic orbit.

5.4 The Kaplan–Yorke conjecture

To calculate the Hausdorff dimension on a computer, or more commonly the capacity (See chapter 2), is even more time and much more memory consuming than to calculate the Lyapunov exponents. On the other hand one may intuitively accept that there is a strong connection. For instance, when all LEs are negative the phase space is contracting in all directions and the attractors must be points of dimension zero. When $\lambda_1 = 0$ and all other exponents are negative it seems reasonable that the dimension must be one.

Kaplan and Yorke have conjectured the following connection between dimension and the LE spectrum:

$$d = N + \left(\sum_{i=1}^{N} \lambda_i \right) |\lambda_{N+1}|^{-1} \tag{5.27}$$

where N is determined by $\sum_{i=1}^{N} \lambda_i > 0$, but $\sum_{i=1}^{N+1} \lambda_i < 0$.

There are counter examples to the relation equation (5.27). In fact, we know one already. From section 3.12 we know that at the period doubling

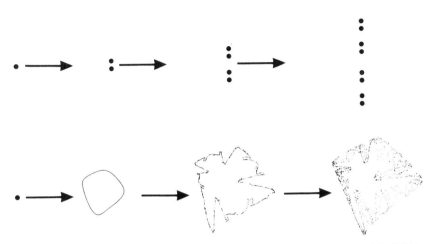

Figure 5.1 The Hopf bifurcation, lower part, compared to the pitchfork bifurcation, schematically.

accumulation points the LE is zero and the dimension of the attractor is $0.538\ldots$, while the Kaplan–Yorke formula gives $d = 1$. However, it is believed that all such examples have zero measure in parameter space, i.e. change a parameter an infinitesimal amount and the formula will be found to be correct.

5.5 The Hopf bifurcation

From the logistic map, we learned that a fixed point may lose stability by splitting into a pair of points constituting the points of a period two orbit. There is another common way for a point to become unstable, namely by turning into a small circle which then increases in size, and at the same time the circle gets deformed as the controlling parameter is increased. This is a *Hopf bifurcation* (see figure 5.1). The motion on the circle is quasiperiodic. Besides period doubling bifurcations, Hopf bifurcations have also been seen in Rayleigh–Benard experiments.

6

DISSIPATIVE MAPS IN HIGHER DIMENSIONS

6.1 The Hénon map

Consider the quadratic map

$$y_{n+1} + B y_{n-1} = 2c y_n + 2y_n^2.$$ (6.1)

For $|B| < 1$ this is Hénon's dissipative map. For $B = 0$ the map is evidently just the logistic map, while for $B = 1$ the map is area conserving. This is seen by writing equation (6.1) in an equivalent form:

$$
\begin{aligned}
x_{n+1} &= y_n + 1 - a x_n^2 \\
y_{n+1} &= b x_n.
\end{aligned}
$$ (6.2)

The Jacobian matrix of this map is then given by

$$j = \begin{pmatrix} -2ax_n & 1 \\ b & 0 \end{pmatrix}$$ (6.3)

and the Jacobian is just $-b$. Since this is a constant we may use equation (5.24) to find the following relation between the two LEs:

$$\lambda_1 + \lambda_2 = \ln |b|.$$ (6.4)

Consequently it is not necessary to calculate more than the largest LE in this case. The eigenvalue equation for the LEs is evidently a quadratic equation which may have either two real and different solutions, or two solutions that are complex conjugates of each other. In figure 6.1 the largest LE is shown for fixed b. As may be seen, the dips in the exponent are no longer infinitely deep like in the logistic map. The characteristic flat regions are due to inequality (5.26) becoming an equality with the result $\lambda_1 = \lambda_2 = \ln(0.3)/2 = -0.602\ldots$. From figure 6.1 one may see that there are period doubling sequences very much as in the case of the logistic map. Since at least one LE is negative the fixed points may be stable, i.e. have only a stable manifold (this is a stable node), or it may have a stable one-dimensional manifold plus an unstable one-dimensional manifold (this is a

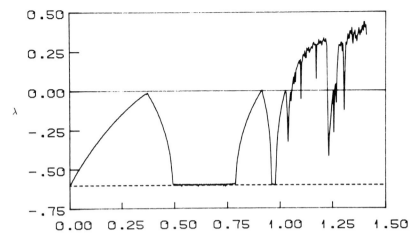

Figure 6.1 The largest Lyapunov exponent for the Hénon map for $b = 0.3$.

hyperbolic point or a saddle point). It is an easy matter to find the fixed points. They are given by

$$x_{0,1} = (-1 + b \pm \sqrt{(1-b)^2 + 4a})/2a. \qquad (6.5)$$

These are points in real space provided

$$a \geq -(1-b)^2/4. \qquad (6.6)$$

Calling the eigenvalues of the Jacobian matrix Λ, the eigenvalue equation is given by

$$\Lambda^2 + 2ax\Lambda - b = 0. \qquad (6.7)$$

The fixed point becomes unstable if $|\Lambda| > 1$. Taking Λ to be on the unit circle in the complex plane, i.e. $\Lambda = e^{i\varphi}$, equation (6.7) splits into two equations

$$\cos 2\varphi + 2ax \cos \varphi - b = 0 \qquad (6.8)$$

and

$$\sin 2\varphi + 2ax \sin \varphi = 0. \qquad (6.9)$$

The last of these equations has one solution $\sin \varphi = 0$, $\cos \varphi = \pm 1$. Inserting this into equation (6.8) shows that there is no solution for the positive sign except at

$$a = -(1-b)^2/4 \qquad (6.10)$$

where the two fixed points coalesce. The negative sign gives $\Lambda = -1$ which corresponds to a period doubling bifurcation. From equation (6.7) one finds that (x_0, y_0) becomes unstable for

$$a > 3(1-b)^2/4 \qquad (6.11)$$

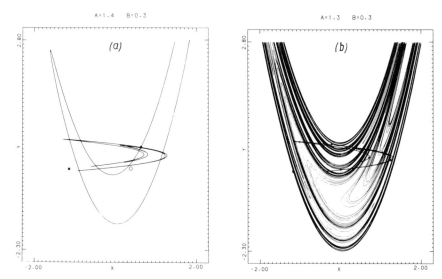

Figure 6.2 (a) Stable and unstable manifolds of the Hénon map. Homoclinic points at A, B, C, D, ... (b) The same as (a), but now the stable manifold has been plotted backwards in time much longer. From Franceschini V and Russo L 1981 *J. Stat. Phys.* **25** 757. Reproduced with permission.

and (x_1, y_1) is always unstable.

Equation (6.9) has also the solution $\cos \varphi = -ax$. Putting this into equation (6.8) gives $b = -1$. This value of b corresponds to a conservative map with properties very different from the dissipative map that we want to study.

Solving the map equations for x_n and y_n expressed in terms of x_{n+1} and y_{n+1}, one sees that the map is *invertible*. Consequently one may use the procedures outlined in section 5.2 to make a graphical representation of the stable and unstable manifolds at parameter values where the motion is chaotic. An example is shown in figure 6.2.

One may also perform an apparently totally different numerical experiment by starting the iteration at an arbitrary point, iterate for say 1000 iterations without recording the results, and then start plotting. The result will be a figure that looks completely identical to the unstable manifold. Figure 6.3 shows the result of such an experiment. Notice the self-similar and Cantor set like structure in the direction normal to the unstable manifold, i.e. in the direction of the stable manifold. Obviously the orbit will, after a sufficiently long time, come arbitrarily close to the unstable manifold. Consequently the unstable manifold may be considered to be identical to the strange attractor. The eigenvalue of the Jacobian corresponding to the stable direction is about 0.1559. We therefore expect that this number

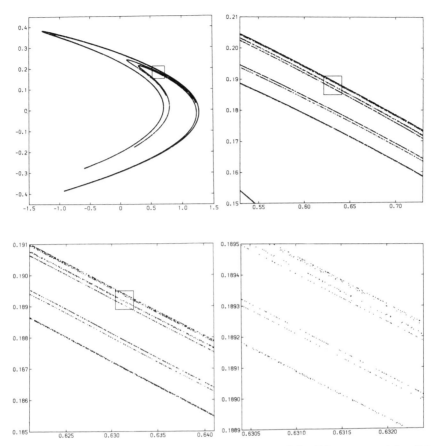

Figure 6.3 The orbit along the strange attractor of the Hénon map at $b = 0.3$ and $a = 1.4$ with an increasing degree of magnification around the fixed point $(x, y) = (0.631\,35\ldots, 0.189\,40\ldots)$. The lines are in the direction of the unstable manifold. From Hénon M 1976 *Commun. Math. Phys.* **50** 69. Reproduced with permission.

shows up as a ratio between the sizes of self-similar structures in the direction normal to the unstable manifold. Measuring on the figures indeed confirms this as well as one can expect.

It was a trivial matter to find the 'orbit' corresponding to the fixed point, and the corresponding Lyapunov exponent may also easily be calculated. Now, at the parameter point corresponding to figure 6.3 the fixed point is unstable, and the largest LE of this 'fixed point orbit' must therefore be bigger than the one corresponding to the strange attractor of figure 6.3, so using the two sets of LEs in the Kaplan–Yorke formula will give different results for the dimension. It is implicit in the concept of dimension that one should get the same result for the dimension independent of what part of the attractor one chooses to look at since the result is supposed to be independent of scale. So, if we find the capacity dimension in the vicinity of the fixed point we must get the same result as in the vicinity of any other point *on the attractor*. But in a small enough neighbourhood of the fixed point the motion may be linearized and should therefore be given by the LEs corresponding to the fixed point. What is the solution to this apparent contradiction? The answer is that the orbit moves out of the linear regime, and then after a certain time gets reinjected into the same regime, carrying along information about all other parts of the attractor. Thus the density of points becomes different from what one gets from linearizing.

Just for the record: the dimension of the strange attractor of figure 6.3 is about 1.26 (using the LEs of the fixed point gives 1.35), a result one may find from box counting algorithms or calculating LEs and using the Kaplan–Yorke conjecture. However, the computing time is claimed to be a factor of 16 less in the last case. In three or higher dimensions the difference is far more dramatic.

6.2 The complex logistic map

One way to get a two-dimensional map from a one-dimensional map is simply to make it complex. The canonical example is again the logistic map, although some of its properties are in some sense not typical of two-dimensional maps.

One may perform a very simple numerical experiment, first performed by Mandelbrot, to see the gross features of the map.

Let the map be given by

$$z \leftarrow z^2 - \mu \qquad (6.12)$$

where z and the parameter μ are complex. The critical point of the map is at $z = 0$. Choose a point in the complex μ plane and start the iteration at $z = 0$. If the orbit does not diverge, a dot is plotted at the corresponding position in the μ plane. By going systematically through the μ plane the resulting plot is the famous Mandelbrot cactus (figure 6.4), and the set of points is called the *Mandelbrot set*. This is the set of values of μ for which the iterative process does not diverge. A physicist might call this a *phase diagram* of the map.

There is a very large literature, some of which is beautifully illustrated, on the complex logistic map. In this short section we shall only seek to understand some of its simplest properties.

Writing the map in its two-dimensional form using

$$z = x + iy \qquad (6.13)$$

Figure 6.4 The Mandelbrot cactus. The black regions are parameter points where the iteration, starting at the origin, never diverges. The lower right cactus is an enlargement of the speck to the right of the top cactus, and the lower left cactus is an enlargement of a tiny spot below the top cactus. From Mandelbrot B B 1983 *Physica* **7D** 224. Reproduced with permission.

we find that the Jacobian matrix of one iteration is given by

$$j = \begin{pmatrix} 2x & -2y \\ 2y & 2x \end{pmatrix}. \tag{6.14}$$

The eigenvalues are $2z$ and $2z^*$. Thus, since the real parts of the eigenvalues are the same, there is no obviously preferred direction, and this is one property that makes this map special. In fact, it is easy to show that this property is true for any orbit of finite length.

The fixed points of the map are

$$z_0 = \tfrac{1}{2} \pm \sqrt{\tfrac{1}{4} + \mu}. \tag{6.15}$$

The fixed points become unstable when $|\Lambda| = 1$, so in the μ plane the fixed points lose stability along the line given by $2z_0 = e^{i\varphi}$ where $\varphi \in [0, 2\pi]$. A

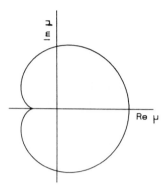

Figure 6.5 The line given by equation (6.16) in the μ plane where the fixed point loses stability.

simple calculation gives

$$\mathrm{Re}\,\mu = (\cos 2\varphi - 2\cos \varphi)/4$$
$$\mathrm{Im}\,\mu = (\sin 2\varphi - 2\sin \varphi)/4. \tag{6.16}$$

The curve is shown in figure 6.5, and as can be seen it exhibits the characteristic shape of the 'trunk' of the full cactus shown in figure 6.4.

To better understand the nature of the branches of the cactus, let us look at values of $\varphi = \varphi_{k,N} \equiv 2\pi k/N$ where $0 < k < N$ and k and N are relative primes. Since $\Lambda^N = 1$ it means that there exists a period N orbit an infinitesimal distance away from the stability region of the period one orbit at the parameter point characterized by $\varphi_{k,N}$. The smallest possible N value is 2, in which case the only possible k is 1 giving $\Lambda = -1$ and $\mathrm{Im}\,\mu = 0$. This is nothing but the period doubling bifurcation, well known from the real case. For $N > 2$

$$\varphi_{k,N} = -\varphi_{N-k,N} \mod 2\pi. \tag{6.17}$$

Thus the branches come in pairs.

If we consider orbits of parameters inside a given branch attached to the trunk at $z_0(\varphi = \varphi_{k,N})$, the orbit is attracted to a period N orbit. Phrased differently the fixed point undergoes a period N-tupling bifurcation at the point in question. The orbits at the conjugated branch are very similar except that the points on the attractor are visited in reversed order. The ratio k/N is again the *winding number*.

Suppose $\mu = 0$ so that the map is $z \leftarrow z^2$. This map has an attracting fixed point at $z = 0$ and a repelling fixed point at $z = 1$. Obviously any orbit that starts out with $|z| < 1$ will converge to the stable fixed point, while orbits with $|z| > 1$ will diverge. On the other hand points

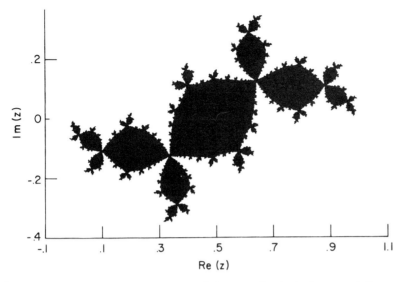

Figure 6.6 Interior of the Julia set of the complex logistic map at the super-stable period three point. From Cvitanovic P and Myrheim J 1989 *Commun. Math. Phys.* **121** 225. Reproduced with permission.

on the unit circle will only map into other points on the unit circle which therefore is an (unstable) *invariant set*. Such an invariant set is called a *Julia set*. For other values of μ the Julia sets are in general not smooth curves, but rather they may be complicated fractal objects as indicated in figures 6.6 and 6.7. One may also regard the Julia set as *the basin of attraction boundary* since by definition these points will not end up on any attractor. (In this connection we may also regard infinity as an attractor.) All images and pre-images of all points in the set are also in the set.

6.3 Two-dimensional coupled logistic map

The Hénon map discussed in section 6.1 was an invertible two-dimensional map. In the following we shall study a non-invertible (an endomorphism) two-dimensional map. It is a special case of the n-dimensional linearly coupled logistic map. In the literature the latter appears as the name of somewhat different maps. One which has been widely used is

$$x^i_{t+1} = (1-\epsilon)f(x^i_t) + \frac{\epsilon}{2}(f(x^{i+1}_t) + f(x^{i-1}_t)) \quad i = 1, 2, \cdots, n. \quad (6.18)$$

Here x^i_t is the ith component of an n-dimensional vector $\boldsymbol{x}_t$ and the function $f(x)$ is the one-parameter function used in the study of the one-dimensional

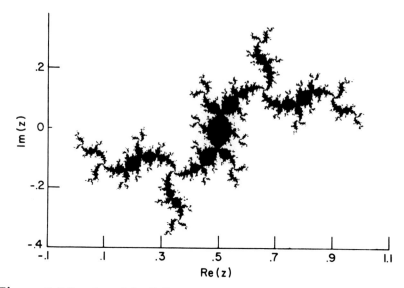

Figure 6.7 Interior of the Julia set of the complex logistic map at the super-stable period nine point. From Cvitanovic P and Myrheim J 1989 *Commun. Math. Phys.* **121** 225. Reproduced with permission.

map. Often cyclic boundary conditions are used. Effectively this means $x_t^{n+1} = x_t^1$ at all times t.

If we consider $f(x)$ as describing the dynamics of a nonlinear oscillator, the second parameter, ϵ, regulates the strength of the coupling between n identical oscillators. When $\epsilon = 0$ the oscillators are decoupled from another. Furthermore the oscillators are coupled directly only to its *nearest neighbours*. The map equation (6.18) is appealing in describing some physical systems. However, nonlinearities enter in several terms, and it is difficult to make exact calculations with this map. Instead we shall consider the simpler, linearly coupled logistic map. It has the form

$$x_{t+1}^i = d(x_t^{i+1} + x_t^{i-1}) + f(x_t^i). \qquad (6.19)$$

This map too has nearest-neighbour couplings. The coupling strength is regulated by the parameter d. Using cyclic boundary conditions, putting $n = 2$ and defining $x_t^1 = x_t$ and $x_t^2 = y_t$, equation (6.19) takes the form

$$
\begin{aligned}
x_{t+1} &= 2dy_t + f(x_t) \\
y_{t+1} &= 2dx_t + f(y_t).
\end{aligned}
\qquad (6.20)
$$

The Jacobian is given by

$$j = \begin{vmatrix} f'(x_t) & 2d \\ 2d & f'(y_t) \end{vmatrix}. \qquad (6.21)$$

Unlike in the case of the Hénon map the Jacobian is not a constant.

6.3.1 The fixed points

The fixed points (x, y) are given by

$$
\begin{aligned}
x &= 2dy + f(x) \\
y &= 2dx + f(y).
\end{aligned}
\tag{6.22}
$$

This set of equations normally has four different solutions. Two of the solutions have $x \neq y$. These fixed points are unstable, but the proof is left to the reader as an exercise.

We now choose

$$
f(x) = 2cx + 2x^2.
\tag{6.23}
$$

This choice is made because in the one-dimensional case the origin is then a stable fixed point for $c \in \langle -\frac{1}{2}, 0]$. Solving equation (6.22) with the condition $x = y$ gives the two solutions

$$
\begin{aligned}
x &= y = 0 \\
x &= y = \tfrac{1}{2} - c - d.
\end{aligned}
\tag{6.24}
$$

Before proceeding, we note that the map is invariant under the transformation

$$
\begin{aligned}
c &\rightarrow 1 - 2d - c \\
(x_t, y_t) &\rightarrow (x_t + d + c - \tfrac{1}{2}, y_t + d + c - \tfrac{1}{2}).
\end{aligned}
\tag{6.25}
$$

In the (c, d) plane there is a line $c = -d + \frac{1}{2}$ which transforms into itself under the transformation (6.25). Thus it is only necessary to investigate one side of this line, and we choose to investigate the side where the origin is a stable fixed point. The transformation (6.25) transforms the second fixed point of equation (6.24) into the origin.

The Jacobian matrix of the origin is

$$
j = \begin{pmatrix} 2c & 2d \\ 2d & 2c \end{pmatrix}.
\tag{6.26}
$$

The eigenvalues are

$$
\lambda_{1,2} = 2(c \pm d).
\tag{6.27}
$$

The fixed point becomes unstable when $|\lambda_{1,2}| = 1$. Since c and d are real, $\lambda_{1,2} = \pm 1$ are the only possible values. Equation (6.27) gives two lines for each $\lambda_{1,2}$ value, so the stable region of the fixed point is a quadrate in the (c, d) plane (see figure 6.8). The two lines corresponding to $\lambda_{1,2} = -1$ are period doubling bifurcation lines. The lines corresponding to $\lambda_{1,2} = 1$ signal a change in symmetry.

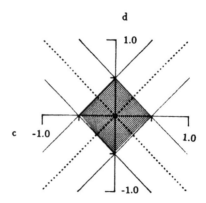

Figure 6.8 The dark square shows the stable region of the origin which is the stable fixed point. Broken lines show where one of the eigenvalues of the Jacobian matrix is zero. From Hansen K T 1988 *University of Oslo Thesis*. Reproduced with permission.

6.3.2 *The stable and unstable manifolds. The structure of the basin of attraction*

The tangent vectors of the stable and unstable manifolds of the fixed point computed as the eigenvectors of the Jacobian matrix equation (6.26) are (not normalized)

$$\left(\begin{array}{c} u \\ v \end{array} \right)_{1,2} = \left(\begin{array}{c} 1 \\ \pm 1 \end{array} \right).$$ (6.28)

The first of these vectors falls along the diagonal. In the region where $|2(c + d)| < 1$, it is evident that part of the diagonal, D, must be part of the stable manifold, S, of the fixed point, so in this case the tangent space is tangent to S, in a finite interval. Since the map is not invertible there is more than one pre-image to an arbitrary point on the diagonal. To see where these pre-images are located, we may subtract the two equations (6.20). The right-hand side is factorizable, and we get

$$x_{t+1} - y_{t+1} = (x_t - y_t)(-2d + 2c + 2x_t + 2y_t).$$ (6.29)

For a point to end up on the diagonal, the left-hand side of this equation must be zero. The right-hand side is zero for points on the diagonal, D, and for points on the transverse line, T, given by

$$x + y + c - d = 0.$$ (6.30)

It is quite easy to go one step further back and find where the pre-images of T are located. Adding the two equations (6.20) and using the condition equation (6.30) on the left-hand side gives after some slight rearrangement

$$[x + (d + c)/2]^2 + [y + (d + c)/2]^2 = [d - c + (d + c)^2]/2.$$ (6.31)

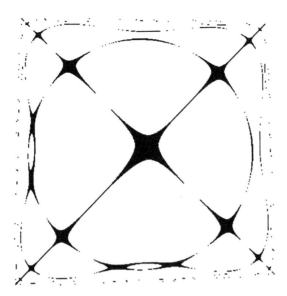

Figure 6.9 Black shows basin of attraction for the period two in-phase attractor at $c = -0.605$, $d = 0.0485$. From Hansen K T 1988 *University of Oslo Thesis.* Reproduced with permission.

This is a circle, C, with its centre at $(-(c+d)/2, -(c+d)/2)$ and radius

$$R = \sqrt{[d - c + (d + c)^2]/2}. \qquad (6.32)$$

Not all of D, T and C belong to S, but we shall not enter into a discussion of that point here. The interval on D that belongs to S is easy to find and is left to the reader as an exercise.

Continuing the iterations backwards in time generates S, and it is quite obvious that its geometry becomes very complicated. The structure of the basin of attraction is strongly influenced by the geometry of S even if the attractor is not the fixed point. This is demonstrated in figure 6.9 where black indicates a basin of attraction, i.e. points where the iterations bring the system to within a small circle around the attractor which in this case is a period two orbit. We shall return to this point in subsection 6.3.4.

6.3.3 Period doubling and Hopf bifurcations

Suppose that the two oscillators are completely decoupled from one another, and that both are in the same period two state. The oscillators may then oscillate *in phase* or *out of phase*. Let us name A and B the two possible values of x_t and y_t. The in-phase mode means that the oscillators will have the sequence $(A, A) \rightarrow (B, B) \rightarrow (A, A)$. The out-of-phase mode

has the sequence $(A, B) \rightarrow (B, A) \rightarrow (A, B)$. These two different period
two modes will be possible also when d is non-zero, but their regions of
stability will not be the same, although there will be an overlapping region
where both modes are possible and stable. In addition their bifurcation
patterns will be different.

The amplitudes of the in-phase period two amplitudes are easy to find,
they are just a repetition of the calculation for the one-dimensional map.
The result is

$$\left. \begin{array}{c} A \\ B \end{array} \right\} = -\tfrac{1}{2}\left(c + d + \tfrac{1}{2}\right) \pm \sqrt{\left(c + d + \tfrac{1}{2}\right)\left(c + d - \tfrac{3}{2}\right)}. \tag{6.33}$$

In order to investigate the stability of in-phase period two oscillations
one must check the eigenvalues of the Jacobian matrix of the map that
takes the system two time steps forwards. This Jacobian matrix is the
product of two local Jacobian matrices of the map

$$\begin{pmatrix} f'(A) & 2d \\ 2d & f'(A) \end{pmatrix} \begin{pmatrix} f'(B) & 2d \\ 2d & f'(B) \end{pmatrix}$$
$$= \begin{pmatrix} f'(A)f'(B) + 4d^2 & 2d(f'(A) + f'(B)) \\ 2d(f'(A) + f'(B)) & f'(A)f'(B) + 4d^2 \end{pmatrix}. \tag{6.34}$$

The eigenvalues are

$$\lambda_{1,2} = f'(A)f'(B) + 4d^2 \pm 2d(f'(A) + f'(B)). \tag{6.35}$$

It is clear that the eigenvalues may only be real. Inserting equation (6.33)
into equation (6.35) one obtains

$$\begin{array}{rcl} \lambda_1 & = & -4(c^2 + d^2 + 2cd - c - d - 1) \\ \lambda_2 & = & -4(c^2 - 3d^2 + 2cd - c - 3d - 1). \end{array} \tag{6.36}$$

Putting $\lambda_{1,2} = \pm 1$ into equation (6.36) gives four different curves in the
(c, d) plane enclosing the stable region of the period two in-phase oscillations. In particular, $\lambda_1 = 1$ gives

$$c + d + \tfrac{1}{2} = 0 \tag{6.37}$$

which is one of the bifurcation lines found earlier for the period doubling
bifurcation of the fixed point. The other bifurcation lines are also easy to
find by putting $\lambda_1 = -1$ and $\lambda_2 = \pm 1$ in equation (6.36).

To find the out-of-phase orbit we have to solve

$$\begin{array}{rcl} B & = & 2dB + 2cA + 2A^2 \\ A & = & 2dA + 2cB + 2B^2. \end{array} \tag{6.38}$$

Subtracting and dividing out the $A = B$ solution gives a linear relation between A and B, and thus a quadratic equation results. Its solutions are

$$\left. \begin{array}{r} A \\ B \end{array} \right\} = -\tfrac{1}{2}(c + d + \tfrac{1}{2}) \pm \tfrac{1}{2}\sqrt{(c - d + \tfrac{1}{2})(c + 3d - \tfrac{3}{2})}. \qquad (6.39)$$

Note that A and B are solutions of the same quadratic equation. Hence when one later needs the combinations $A + B$ and AB they may be read off directly as the negative coefficient of the first degree term and the constant term respectively.

To check the stability of the out-of-phase orbit we must find the eigenvalues of the Jacobian matrix

$$\begin{pmatrix} f'(A) & 2d \\ 2d & f'(B) \end{pmatrix} \begin{pmatrix} f'(B) & 2d \\ 2d & f'(A) \end{pmatrix}$$
$$= \begin{pmatrix} f'(A)f'(B) + 4d^2 & 4df'(A) \\ 4df'(B) & f'(A)f'(B) + 4d^2 \end{pmatrix}. (6.40)$$

Its eigenvalues are

$$\lambda_{1,2} = f'(A)f'(B) + 4d^2 \pm 4d\sqrt{f'(A)f'(B)}. \qquad (6.41)$$

Since $f'(A)f'(B)$ may become negative, $\lambda_{1,2}$ may be complex. If that is the case, the eigenvalues are the complex conjugates of each other. Consequently when $|\lambda_1| = 1$ then also $|\lambda_2| = 1$. Equation (6.41) may be solved with respect to $f'(A)f'(B)$. Dropping the index on λ one finds

$$f'(A)f'(B) = \lambda + 4d^2 \pm 4d\sqrt{\lambda}. \qquad (6.42)$$

Assume that the eigenvalues are complex. Since the left-hand side of equation (6.42) is real, the imaginary terms on the right-hand side must cancel. Thus with $\lambda = \exp(i2\pi k/N)$ and with k and N relative primes it follows that

$$d = \tfrac{1}{2}\cos(\pi k/N) \quad (\lambda_{1,2} \neq 1). \qquad (6.43)$$

Inserting this back into equation (6.42) gives

$$f'(A)f'(B) = -1 + 4d^2 \qquad (6.44)$$

and inserting the explicit values for $f'(A)$ and $f'(B)$ into equation (6.44) gives a line in the parameter plane where the out-of-phase period two oscillation goes through a Hopf bifurcation. Its position is given by

$$c = \tfrac{1}{2} - d - 2\sqrt{(\tfrac{1}{2} - d)(\tfrac{3}{4} - d)}. \qquad (6.45)$$

This line is shown as line β_2 in figure 6.10. In particular if we pick integers k and N, equations (6.43) and (6.45) give the start point of an Arnol'd tongue

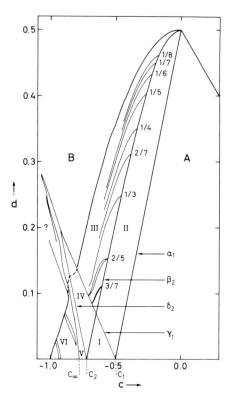

Figure 6.10 An overview of the parameter plane for positive d. Region A: the origin is stable. Region B: all starting points lead to diverging orbits. Region I: in-phase and out-of-phase period two orbits are both stable. Region II: only out-of-phase period two is a stable orbit. Region III: chaotic and quasiperiodic orbits. Arnol'd tongues are indicated. Region IV: the in-phase period two is stable. From Frøyland J 1982 *Physica* **8D** 423. Reproduced with permission.

of winding number k/N [1]. The periodicity of the orbit is $2N$ since it is a period two that bifurcates. Note that there is a region in parameter space where both in phase and out-of-phase oscillations have stable attractors. See figure 6.10.

In addition to the period two orbits given by equation (6.33) and equation (6.39) there are other unstable period two orbits that may be found by solving the complete set of period two equations.

Why do the in-phase orbits only have period doubling bifurcations while the out-of-phase orbits may have Hopf bifurcations? For an orbit to have a Hopf bifurcation it is necessary for its Jacobian matrix to have complex

---

[1] We have not given a general definition of winding number of a two-dimensional map, and shall only ascribe a winding number to Arnol'd tongues.

eigenvalues. If we look at the matrix of equation (6.34), it is symmetric and thus may only have real eigenvalues, while the matrix of equation (6.40) is not symmetric and therefore in general may have complex eigenvalues.

6.3.4 Basins of attraction

Assume again that the oscillators are completely decoupled, and that each oscillates in a stable period N mode. Depending on the relative phase of the two oscillators there will be N different period N modes of the whole system. Each of these N stable orbits will have its own basin of attraction. These basins of attraction will in general be interwoven in a complicated pattern.

We shall consider the simplest non-trivial example of the stable period two orbits and see how its basin of attraction changes when the coupling is strengthened, i.e. when d is increased from zero. Figure 6.11 shows the structure of the basin of attraction at $d = 0$. It consists of rectangles that become infinitely narrow as the outer basin boundary is approached. The fixed point and all its pre-images are on the basin of attraction boundary. How this structure changes depends on what direction one moves in the parameter plane. Keeping c fixed and increasing d slightly in the positive direction while staying inside the region where both in-phase and out-of-phase oscillations are stable (region I of figure 6.10), the squares of the out-of-phase basin of attraction get their corners rounded (see plate 6). If we move in the direction of the line $c + d + \frac{1}{2} = 0$ where the in-phase orbit no longer exists, the corresponding basin of attraction becomes smaller and smaller (see figure 6.9).

Within region II of figure 6.10 the out-of-phase period two orbit is the only stable attractor. For this attractor the stable manifold of the fixed point act as a repellor. This is illustrated in plate 7. Compare this picture to figure 6.9 where the in-phase orbit period two orbit still exists.

By definition, iterating a point on the basin boundary gives an orbit that does not have any of the attractors as a limiting point. Thus the entire orbit has to stay on the boundary. The boundary is therefore an invariant Julia set. It was mentioned previously that there exist period two orbits other than those given by equations (6.33) and (6.39). These are found on the boundary, and in fact the entire boundary may be considered to consist of the fixed point, the pre-images of these unstable period two points, their stable manifolds and pre-images.

In the region of the break up of the Arnol'd tongues there is generally more than one attractor. Their basins of attraction are complicated fractals (see plate 8) whose general structures may be qualitatively understood from analysis of the stable and unstable manifolds of the fixed point. The unstable manifold must make an infinity of bends. At some parameter point a bend may be tangent to the stable manifold. Since the tangent point lies

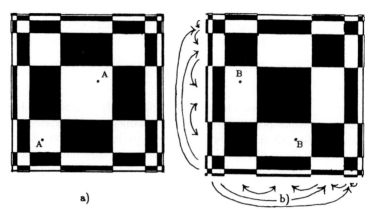

Figure 6.11 White rectangles indicate basin of attraction at $d = 0$ for (a) the in-phase period two orbit (points A) and for (b) the out-of-phase period two orbit (points B). From Hansen K T 1988 *University of Oslo Thesis*. Reproduced with permission.

on both the stable and the unstable manifold, it is a homoclinic point, and the phenomenon is called a *homoclinic tangency*. Such tangencies play a crucial role in understanding the nature of high period orbits, but a discussion of homoclinic tangencies is outside the scope of this book.

6.3.5 Lyapunov exponents

The system is two dimensional, so there are two Lyapunov exponents for each attractor. They may be either different, $\lambda_1 > \lambda_2$, or a degenerate pair, $\lambda_1 = \lambda_2$, as explained in section 5.3. This may be seen in figure 6.12 where the Lyapunov exponents have been calculated for $d = 0.3$. There is a bifurcation each time the biggest Lyapunov exponent touches the zero line.

Following the exponents in figure 6.12 from right to left, there is first a stable fixed point with two different exponents, then there is a period doubling bifurcation at $c = -0.2$. The resulting period two is an out-of-phase oscillation. Further to the left there is a collapse of exponents and this collapse lasts till the Hopf bifurcation at $c = -0.4$. After the Hopf bifurcation, the biggest exponent is mainly zero within the precision of the calculation, but at least one fairly wide dip may be seen. Just as for the circle map, such dips are the result of crossing Arnol'd tongues. With better resolution one would have seen many more such dips. The fact that one exponent may be zero and the other one negative means that there is a direction 'along' the orbit which is not contracting, while the tranverse direction shrinks. The result is that the attractor exists on two topological

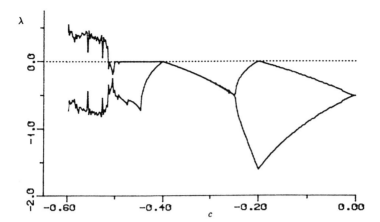

Figure 6.12 The two Lyapunov exponents as a function of c for $d = 0.3$. Note the collapse of exponents and the Hopf bifurcation. The dip in the biggest exponent at about $c = -0.5$ is due to a crossing of an Arnol'd tongue. From Hansen K T 1988 *University of Oslo Thesis.* Reproduced with permission.

circles (two because the Hopf bifurcation starts from a period two orbit). The dimension of the attractor is then exactly one. The circles break up into N points whenever an Arnol'd tongue of winding number k/N is crossed. As one makes c more and more negative the 'circle' becomes more and more wavy, and at some point the 'circle' grows small cusps. By then the attractor has become a fractal of dimension bigger than one, and the biggest exponent is bigger than zero (see figures 6.12 and 6.13).

The c value where the biggest exponent makes the transition from being zero to a positive value corresponds to the critical value $K = 1$ in the circle map. Varying d, one may imagine that there is a whole critical line, as in the case of the circle map. However, the precise position of such a line is not easy to find because Arnol'd tongues in general cut into the chaotic region (positive biggest Lyapunov exponent), and unlike the case of the circle map we have no analytic expression for the critical line. When c is decreased further the motion becomes more and more chaotic as the biggest exponent increases, although this increase is not monotonic and, similar to what is observed for the logistic map, the chaotic region is intersected by stable periodic windows. Most of these do not have their origin as Arnol'd tongues. Instead they are very intimately connected to the behaviour of the stable and unstable manifolds of the fixed point.

Because there may be more than one stable attractor at a given parameter point, there may be more than one set of Lyapunov exponents. All starting points within one basin of attraction give the same result, but keeping the starting point fixed as parameters are varied may cause basin boundaries to be crossed resulting in sudden jumps in the Lyapunov

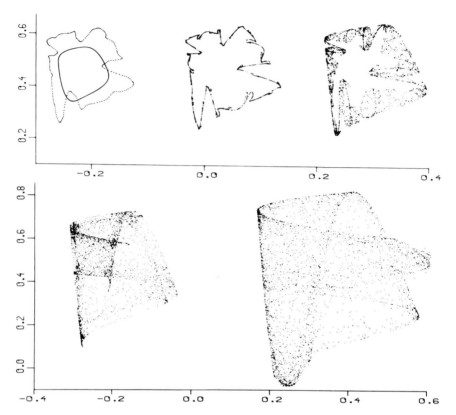

Figure 6.13 Plot of x_t against y_t at $d = 0.06$, only each other point is plotted due to the symmetry. Starting values $(x_1, y_1) = (0.05, -0.02)$. In the upper part of the figure $c = -0.68, -0.705, -0.710$ and -0.715. The middle and rightmost orbits have been displaced 0.25 and 0.50 to the right, respectively. In the lower part of the figure $c = -0.73$ and -0.74. The last plot is displaced 0.5 to the right. From Frøyland J 1982 *Physica* **8D** 423. Reproduced with permission.

exponents. If one wants to avoid or minimize such jumps, one may use 'adiabatic' starting points, meaning that the last phase space point in the iteration at one parameter point is used as the phase space starting point at the next nearby parameter point. A simple algorithm to calculate Lyapunov exponents for systems of dimension up to three is described in appendix 2.

When the map has more than one parameter, it becomes extremely tedious to study the very large number of orbits necessary to gain a reasonable overview of the model. It is somewhat less tedious to make cuts in the parameter plane and calculate the Lyapunov exponents along the cut. The topology of the orbits change only at each bifurcation, i.e. each time

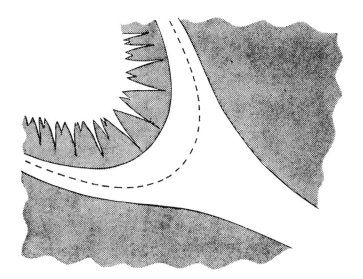

Figure 6.14 Schematic drawing of the Hopf bifurcation pattern of an Arnol'd tongue. From Hansen K T 1988 *University of Oslo Thesis*. Reproduced with permission.

the biggest Lyapunov exponent passes through zero. The human effort may be further reduced if one calculates the Lyapunov exponents on a fine grid (for instance in as many points as there are pixels on a monitor screen) and uses colour codes to indicate the result. A phase diagram constructed in this way is shown in plate 2. Knowing the colour code very quickly gives a good overview of the model. The price is quite long computer times. The complexity of the phase diagram may change greatly from one region of parameter space to another. It may therefore be desirable to magnify some regions. This is in principle trivial, but normally one also has to increase the precision of the calculated exponents. Magnifications of parts of plate 2 are shown in plates 3, 4 and 5.

6.3.6 More about Arnol'd tongues

We saw in chapter 4 on the circle map that Arnol'd tongues split in two at some point above the critical line. Orbits in the tongue would lose stability either by period doublings or by tangent bifurcations into quasiperiodicity. Another characteristic feature was the swallowtail structures. In the case of Arnol'd tongues of the two-dimensional, linearly coupled map, it turns out in numerical experiments that the tongues may break up into new tongues. This process may continue in cascades and seems to be a common way an Arnol'd tongue breaks up on its complicated route to chaos in this

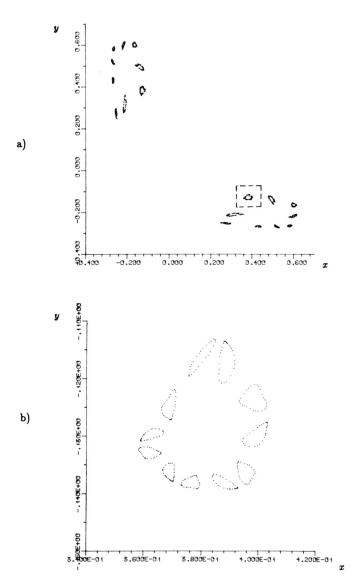

Figure 6.15 Quasiperiodic orbit at $d = 0.08647$ and $c = -0.684105$ after one period doubling bifurcation (two main groups of points in (a)), one Hopf bifurcation of winding number 4/9 (nine smaller groups of points in (a) within each big group), and then a second generation Hopf bifurcation. A magnification of the orbit of (a) inside the stippled square is shown in (b). From Hansen K T 1988 *University of Oslo Thesis*. Reproduced with permission.

particular map (see figure 6.14, figure 6.15 and plate 5). However, some Arnol'd tongues seem to break up according to the same rules as were found for the circle map.

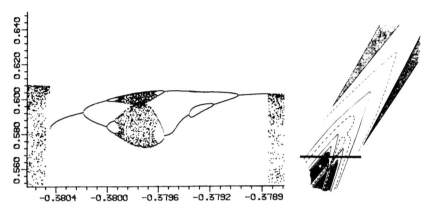

Figure 6.16 To the left x_t is plotted against c at $d = 0.40156$. The Arnol'd tongue has winding number $k/N = 1/11$, and only each eleventh point on the orbits has been plotted. To the right is an overview of the bifurcation lines of the Arnol'd tongue in the vicinity of the cut which is shown as a dark line. From Hansen K T 1988 *University of Oslo Thesis*. Reproduced with permission.

An often used method of display is to plot the time series, or a coarse-grained time series, of one variable against one parameter. In higher dimensional maps such plots may be confusing if one does not make other investigations. One such plot is shown in figure 6.16, together with an overview of the bifurcation pattern that has been crossed.

7

CONSERVATIVE MAPS

Conservative maps are maps where the absolute value of the Jacobian (determinant) is one at every point in phase space. As a consequence any phase space volume element is constant in magnitude as time evolves. In two dimensions the 'volume' is two dimensional, and such maps are often refered to as *area preserving maps*.

In many ways conservative maps are different from dissipative maps. The simplest is the absence of linearly stable fixed points and linearly stable periodic orbits. This is obvious because the Jacobian is the product of all the Jacobian matrix eigenvalues. As a consequence all eigenvalues cannot have an absolute value less than one—a requirement for a stable fixed point. However, quasiperiodicity is possible, and is a very important type of motion.

The most well-known type of conservative systems are Hamiltonian systems. The theory of Hamiltonian systems is a large subject which we shall not enter into. Instead we shall consider a couple of simple examples of conservative maps, and also give an indication of what the KAM theorem is concerned with.

7.1 The twist map

Consider the following linear *twist* map

$$\begin{aligned}
x_{n+1} &= \cos(2\pi\alpha)\,x_n - \sin(2\pi\alpha)\,y_n \\
y_{n+1} &= \sin(2\pi\alpha)\,x_n + \cos(2\pi\alpha)\,y_n
\end{aligned} \tag{7.1}$$

where α is the parameter of the map. Obviously the origin is a fixed point, and it is a quick calculation to show that the eigenvalues of the Jacobian matrix are of the form $e^{\pm i2\pi\alpha}$. Thus the origin is an elliptic fixed point. Since the Jacobian is one, the map is conservative. It is fairly obvious that the map (7.1) would be the result of linearizing a two-dimensional conservative map around one of its elliptic fixed points and moving the origin to the fixed point. In polar coordinates the map (7.1) is just

$$\begin{aligned}
\theta_{n+1} &= \theta_n + 2\pi\alpha \quad \mathrm{mod}\ 2\pi \\
r_{n+1} &= r_n.
\end{aligned} \tag{7.2}$$

Independent of the value of α, which we shall call the winding number, the points on the orbit will be confined to a curve, an *invariant circle*, or in more general cases to an *invariant torus*. The radius of the circle is just r_0. Suppose α is a *rational* number m/n. Then after n iterations one has returned to the starting point, and thus the orbit is periodic of period n. In this case every point on the circle is a periodic point of period n, but one orbit only traces out n points. If on the other hand α is irrational the orbit never returns to the starting point, and the orbit will eventually trace out the whole circle except points of measure zero. The motion takes place on an *irrational torus* as opposed to on a *rational torus* in the first case. The twist map may be considered a mapping in a Poincaré section of the phase space of a Hamiltonian system, and therefore it makes sense to consider also the isolated points of a rational orbit to be an invariant torus.

Let us change the map (7.2) slightly by introducing the *small* perturbing functions f and g in the following way:

$$\begin{aligned}
\theta_{n+1} &= \theta_n + 2\pi\alpha + f(\theta_n, r_n) \\
r_{n+1} &= r_n + g(\theta_n, r_n).
\end{aligned} \tag{7.3}$$

The functions f and g are supposed to be chosen in such a way that the map remains conservative. Suppose that $\alpha = m/n$, so that n iterations of the original map takes the point (r_0, θ_0) back into itself. Starting in that same point the map n iterations of equation (7.3) gives a new point *near* (r_0, θ_0). At least for small perturbations it follows from continuity that it must be possible to find a new starting point (ρ_0, θ_0) near (r_0, θ_0) so that $\theta_n = \theta_0$. Letting θ_0 run through all points on the circle uniquely defines a new closed curve, ρ_0, which is no longer a circle. Let us consider the nth image of ρ_0, ρ_n. In general the two curves will not be the same, but since the map is conservative the areas enclosed by the two curves must be the same. Consequently the curves must cross each other, and since the curves are closed the number of crossings must be an even number. The crossing points are then periodic points, and it can be proved that they are alternate hyperbolic and elliptic. Remember that to each hyperbolic point there are associated stable and unstable manifolds, while every elliptic point has a centre manifold with invariant tori associated with it. The situation is illustrated in figure 7.1.

Suppose the strength of the perturbing functions is increased. The curves ρ_0 and ρ_n may get increasingly wiggly and one may reach a situation where one elliptic crossing point turns (bifurcates) into three periodic points with the original point as a new hyperbolic point, and in addition two new elliptic points, so there are now twice as many of both elliptic and hyperbolic points. The period of the orbit constituted by the elliptic points is now $2n$, i.e. the bifurcation is a period doubling bifurcation. This is illustrated in figure 7.2.

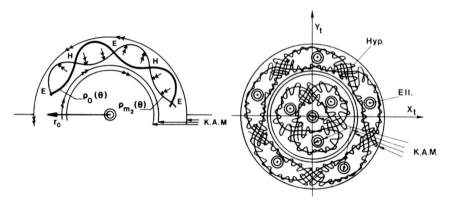

Figure 7.1 Phase space structures near an elliptic origin. (*a*) The curves ρ_0 and ρ_n. (*b*) Hyperbolic and elliptic points and the associated manifolds of the hyperbolic points (separatrices) and some invariant tori on the centre manifolds of the elliptic points. From Helleman R H G 1980 *Fundamental Problems in Statistical Physics V*, ed E G D Cohen (Amsterdam: North-Holland) p 165. Reproduced with permission.

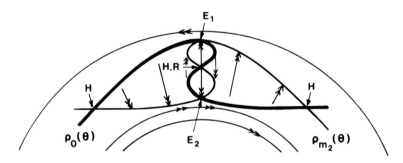

Figure 7.2 Closed curves ρ_0 and ρ_n when the perturbation has become so large that each elliptic point has bifurcated into a hyperbolic and two elliptic points. From Helleman R H G 1980 *Fundamental Problems in Statistical Physics V*, ed E G D Cohen (Amsterdam: North-Holland) p 165. Reproduced with permission.

It should be quite evident that each of the new elliptic points may be made into the origin of a new expansion of the same type as outlined above, so as the nonlinearity increases one expects more and more bifurcations.

Orbits originating in the vicinity of a hyperbolic point will be chaotic, while near elliptic points there will be invariant tori. As the nonlinearity increases we expect the occupied part of phase space to become more and more complicated with large regions of chaotic behaviour, but these regions

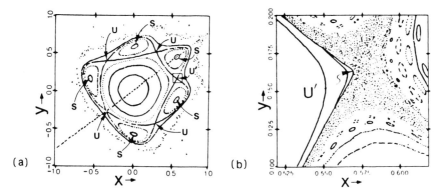

Figure 7.3 See text. (*b*) is a magnification of a small region around one of the hyperbolic points in (*a*). All points of the chaotic region are on the same orbit of a conservative, two-dimensional, quadratic map. From Bountis T and Helleman R H G 1981 *J. Math. Phys.* **22** 1867. Reproduced with permission.

are broken up into larger and smaller islands of periodic or quasiperiodic motion on invariant tori (see figure 7.3).

7.2 The KAM theorem

We saw in the preceding section that the periodic orbit, i.e. the rational tori, could disappear through period doubling bifurcations when the strength of the perturbation was increased. The Kolmogorov, Arnol'd and Moser (KAM) theorem says essentially that only those invariant tori that are *sufficiently irrational* will survive the perturbation with the same winding number. The expression 'sufficiently irrational' is given a more precise meaning. For two-dimensional conservative maps it is given by the condition

$$|\alpha - m/n| \geq \frac{c}{n^\mu} \qquad \text{for all integers } m \text{ and } n \qquad (7.4)$$

where $c > 0$ is a constant which in general is hard to find, and $\mu > 2$ is another constant dependent on both the map and n. However, most irrational numbers do satisfy equation (7.4) and so the corresponding tori do survive. On the other hand, in the vicinity of 'very' rational numbers there are no tori that survive. Thus one may expect gaps in phase space near orbits corresponding to such 'very' rational numbers.

For higher dimensional maps the expression (7.4) is somewhat more complicated due to the fact that there may be many independent frequencies. The proofs of the KAM theorem are complicated and far beyond the scope of this book.

7.3 The rings of Saturn

The rings of Saturn are thought to be essentially 'invariant tori' and the spaces between them KAM gaps (although the full explanation certainly is much more subtle). The rings consist of stones in a thin layer. The essential perturbing force is thought to come from the moon Mimas circling at an average distance of 185.7×10^3 km from the centre of Saturn.

We can make a toy model of the rings of Saturn that displays the essential idea. We start by considering a small object of unit mass moving in a circular orbit around any planet. The angular velocity of the object is ω and its distance from the planet's centre is r. The magnitude of the gravitational force is proportional to $1/r^2$ and equals the centripetal acceleration $\omega^2 r$ per unit mass. From this we find that for all circular orbits $\omega^2 r^3 = $ constant. We choose a reference orbit of radius r_0. The time it takes to make one revolution in that orbit is our unit of time. We call θ_n the angle that the radius vector of the object makes to some fixed reference direction at time n. For the reference orbit θ_n increases by an amount of 2π in each time step, and in general

$$\theta_{n+1} = \theta_n + \frac{2\pi r_0^{3/2}}{r_n^{3/2}} \tag{7.5}$$

for a circular orbit. Although this is a rule for circular orbits, we consider the perturbations to be small so that the orbits are very nearly circular, and equation (7.5) is taken to be the general rule for updating the angle.

We now consider a system that consists of the planet Saturn, its moon Mimas and test particles moving in the space between the moon and the planet's surface. We assume that Mimas is moving in the reference orbit, and that its gravitational field influences any small object circulating around Saturn. We want a rule to update also the radial distance after each new time step. To find such a rule which is in agreement with all celestial mechanical laws is difficult, but we are only looking for a model that displays the KAM gaps approximately at the positions of the largest gaps in the ring system. One indispensable requirement is that the map must be conservative. This requirement makes it somewhat difficult to find a rule of the form $r_{n+1} = f(r_n, \theta_n)$. If we recognize that the perturbation creates accelerations in the radial direction, we may consider maps of the form

$$r_{n+1} - 2r_n + r_{n-1} = f(r_n, \theta_n). \tag{7.6}$$

In infinitely small time steps, the left-hand side of equation (7.6) is proportional to the radial acceleration. In large time steps like ours, we may take the left-hand side to represent some 'average' acceleration over one time step, and thus the right-hand side must represent some average gravitational force felt by the object due to the presence of Mimas. By introducing

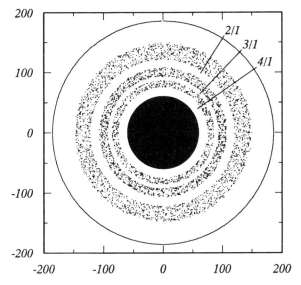

Figure 7.4 Three independent orbits of the toy model of the rings of Saturn (equations (7.5), (7.7) and (7.8)). The axis units are 10^3 km. The reference orbit is $r_0 = 185.7 \times 10^3$ km, and the radius of Saturn is $r_{\text{Saturn}} = 60.4 \times 10^3$ km. The parameter of equation (7.7) $A = 2.0 \times 10^{12}$ km^3. Saturn is shown as a black disc, and the reference orbit of Mimas is shown as a circle.

a new variable $h_n = r_{n-1}$ equations (7.5) and (7.6) together constitute a three-dimensional conservative map independent of the function $f(r_n, \theta_n)$. The proof is left as an exercise.

The function $f(r_n, \theta_n)$ must be periodic in θ_n. Furthermore $f(r_n, \theta_n)$ must be increasing when $|r_n - r_0|$ decreases. Without any further arguments we assume

$$r_{n+1} \qquad 2r_n - h_n + \frac{A \cos \theta_n}{(r_0 - r_n)^2} \qquad (7.7)$$

$$h_{n+1} = r_n \qquad (7.8)$$

for orbits where $r_n - r_0 < 0$. The constant A is a constant that is in principle calculable. It is directly proportional to the mass of Mimas, but we take it as a parameter that to some extent can be used to regulate the widths of the gaps, but the result in the form of figures like figure 7.4 is quite insensitive to the value of A. Figure 7.4 shows three different orbits of the model. If one follows the details of the motion, one sees that there are fairly regular oscillations between certain boundaries in the radial direction. When A is increased the oscillation time within the boundaries decreases, while the boundaries change very little. In figure 7.4 the innermost orbit is started out somewhere around midway between the orbits corresponding to $\omega = 4\omega_{\text{Mimas}}$ and $\omega = 3\omega_{\text{Mimas}}$, or winding numbers 4/1 and 3/1. Correspondingly the middle orbit is started between the orbits of winding number 3/1 and 2/1, and the outermost orbit is started between 2/1 and 1/1. Different starting points in the same region give visually the same picture. Attempts to start orbits near an $n/1$ orbit systematically lead to negative or very large r_n after a few time steps. The radius corresponding to the golden mean winding number is about 134.7×10^3 km—almost in the middle of the outer ring.

To investigate the gaps of the model properly, one should run it with a very large number of different starting points. Indications are that the picture does not change very much.

Reasonable changes in the function $f(r_n, \theta_n)$ of equation (7.6) from what it is in equation (7.7) do not change the general picture of figure 7.4, but of course the relative ratios of the widths of gaps and rings change somewhat given that some parameter is used to approximately reproduce the width of, say, the innermost gap or ring. If the strength of the perturbation is increased, sooner or later the outermost ring will disappear, and if further increased all the rings may disappear.

It is emphasized that the real rings of Saturn show very complicated structures that the toy model does not by any means reproduce.

8

CELLULAR AUTOMATA

In the previous chapters we have been dealing with maps which may be regarded as dynamical systems having a finite number of degrees of feedom (the dimensionality of the map), discrete time, but continuous phase space variables. Systems even simpler than these maps may be formed by letting phase space be discrete also. Such systems are called *cellular automata*. Cellular automata have five characteristics:

- they exist on lattices of discrete sites
- the time evolution takes place in discrete steps
- each site has a finite number of possible states
- each site evolves in time according to a deterministic rule
- the rule for each site depends only on some finite neighbourhood on the lattice and a finite number of previous time steps.

In spite of their simplicity cellular automata show a surprising ability to reproduce complex dynamics. Applications of cellular automata can be found in non-equilibrium physics, hydrodynamics, population dynamics, chemical reactions, epidemiology, parallel computing, geophysics (earthquakes) etc. Because of their discrete nature, cellular automata are particularly well suited for simulations on digital computers. In this chapter we shall only have a brief look at the very simplest of non-trivial cellular automata.

The maps we have studied so far have had a phase space dimension of at most two. In return for discretizing the phase space we shall increase its dimension to some number, typically in the range 50 to a few hundred, in order to discover possible regularities of the system.

A simple cellular automaton may be considered to be a lattice where the sites may have a finite number k of possible values. In the following we shall take the automaton sites to be arranged on a one-dimensional lattice, all sites shall be identical and there is dependence only on the previous time step. In order to further simplify the automaton, we assume only 'nearest-neighbour' interactions. This means that the value at one site depends only upon the value of the site itself and its two immediately adjacent sites in the previous time step. Thus there are k^3 possible initial 'local' states at any given site. Since there are k possible results, the total number of possible rules are k^{k^3}. In the simplest case when $k = 2$ there are 256 possible rules. This number increases extremely rapidly with increasing k. Thus $k = 3$ gives about 7.6×10^{12} possible rules.

Table 8.1 Connection between the local states and the updated central site value.

State	111	110	101	100	011	010	001	000
New value	α_1	α_2	α_3	α_4	α_5	α_6	α_7	α_8

In automata with dependence only on the state in the previous time step we may define a range r so that the dependence is on the $2r$ nearest neighbours in addition to the state of the site itself.

In the following we shall only consider $r = 1$, $k = 2$ automata. Each local state may be characterized by a three digit binary number and the outcome of the updating by a one digit binary number α_n where the index $n = (1, 2, \ldots, 8)$ runs over the eight possible local states. Concatenating these numbers into one eight digit binary number $\alpha_1\alpha_2\alpha_3\alpha_4\alpha_5\alpha_6\alpha_7\alpha_8$ one obtains a unique characterization of the cellular automata. The connection between the local state and α_n is given in table 8.1.

Consider what happens to the 'null' state where all sites have the value zero. According to table 8.1 the result will be a state where all sites have value α_8. If $\alpha_8 = 0$ the null state remains unchanged with time. If $\alpha_8 = 1$ all sites in the next time step will have the value α_1. If $\alpha_1 = 0$ the result is identical to the initial state, and the orbit is a period two oscillation. If $\alpha_1 = 1$ all subsequent time steps will have states where all sites have value 1. We may find it a desirable property of the automaton that the null state is left unchanged with time. This forces $\alpha_8 = 0$. If we consider the system to be homogeneous and isotropic we must make the rules reflection symmetric. This means that the local states 110 and 011 must produce the same result, or $\alpha_5 = \alpha_2$. For the local state 100 to give the same result as 001 we must have $\alpha_7 = \alpha_4$. These constraints leave us with the 'desirable' sets of rules $\alpha_1\alpha_2\alpha_3\alpha_4\alpha_2\alpha_6\alpha_40$. Thus it takes five independent binary numbers to characterize one of our automata and consequently we are left with only 32 automata to investigate.

We start by considering the state with only one site value different from zero. Since $\alpha_8 = 0$, only the site itself and its immediate neighbours can become changed during the first time step. Listing only sites that may have been changed the result is $\alpha_4\alpha_6\alpha_4$. Evidently all subsequent results will be $0\alpha_60$ if $\alpha_4 = 0$. If $\alpha_4 = 1$ the leftmost (rightmost) non-zero site will advance one step to the left (right) in each timestep. This can be seen in figure 8.1.

Rule no. 0 is trivial since it takes any state into the null state after one time step. Rule no. 204 which has the binary representation 11001100

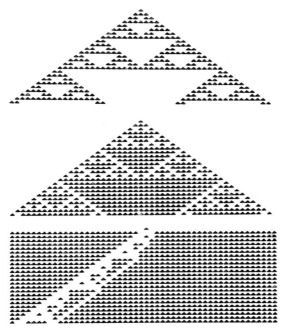

Figure 8.1 Time evolution from a state that contains only one non-zero site. Rule no. 126 at the top, rule no. 182 in the middle and the 'undesirable' rule no. 225 at the bottom. Time starts at the top of each figure. Sites with non-zero value are shown as little triangles.

acts as the identity transformation since it takes any state into itself. The structures of the state evolutions shown in the two top examples of figure 8.1 are very regular and makes it possible to construct the long time behaviour of the cellular automata.

Making these simulations on a computer of the real world, one has to have only a finite number of sites, and therefore one must fix some boundary conditions. There are two common ones. One is to assume the null boundary condition, i.e. that everything outside the boundary is zero. The other one is to have *periodic boundary conditions* in which case the line of sites effectively is a circle so that site number one is a nearest neighbour of site number n in an n-site automaton. Provided the automaton has many more sites than the number of sites in the local neighbourhood (in our case three), the choice of boundary conditions seems to play a minor role.

The initial state of our first experiment was very special. We shall now consider starting from 'random' states. These may be generated by giving a probability, p, that the site value at any given site is one, and then use a random number generator to generate sample random states. Examples of runs starting from random states are shown in figure 8.2. It

Figure 8.2 Time evolution for 200 time steps starting from 100 sites random states with $p = 0.5$. Rule no. 126 to the left, rule no. 225 in the middle, and no rule, i.e. a random sequence of random states, to the right.

is apparent that the state quickly becomes far from random. Instead the automaton organizes the state into structures having different length scales. This ability to *self-organize* is one of the more remarkable properties of the cellular automata. Some understanding of the phenomenon can be gained by considering the result of one iteration on the 16 possible local states containing four nearest neighbours. Each such state uniquely predicts the next updated values of the two middle sites. The general results, and the results in a few sample cases, are given in table 8.2.

For the sake of simplicity, suppose $p = 0.5$. Then all the local states listed in table 8.2 have the same probability. If we count the number of zeros and ones in all the states, there are equally many of each. In 4 out of 16 states the two central sites both have the value one. If we look in table 8.2 we see that for rule no. 90 this ratio is the same also after the updating, while for rule no. 126 10 out of 16 states have site value one for the two central sites. In the last case the probability of having long sequences of ones is much greater than in an initial random state. Using rule no. 126 one sees that sequences more than three sites long of the same site value, in the next step produce a shorter sequence of zeros. This gives rise to the typical triangular structures that may be seen in figure 8.2.

More complicated cellular automata than those studied here may be constructed in many ways, for instance by increasing the dimension of the underlying lattice, increasing the size of the neighbourhood, increasing the number of possible values at one site, having more than one type of site, etc. The number of possible rules becomes so large that it is totally impos-

Table 8.2 Results from updating the two central sites of local four site states.

Local state	1111	1110	1101	1100	1011	1010	1001	1000
General result	$\alpha_1\alpha_1$	$\alpha_1\alpha_2$	$\alpha_2\alpha_3$	$\alpha_2\alpha_4$	$\alpha_3\alpha_2$	$\alpha_3\alpha_6$	$\alpha_4\alpha_7$	$\alpha_4 0$
Rule no. 90	00	01	10	11	01	00	11	10
Rule no. 126	00	01	11	11	11	11	11	10

Local state	0111	0110	0101	0100	0011	0010	0001	0000
General result	$\alpha_5\alpha_1$	$\alpha_5\alpha_2$	$\alpha_6\alpha_3$	$\alpha_6\alpha_4$	$\alpha_7\alpha_5$	$\alpha_7\alpha_6$	$0\alpha_7$	00
Rule no. 90	10	11	00	01	11	10	01	00
Rule no. 126	10	11	11	11	11	11	01	00

sible to investigate all possibilities even for the most modest generalization. Fortunately it appears to be possible to classify the cellular automata into a few classes. In practical cases the number of rules may often be largely constrained by the problem itself.

9

ORDINARY DIFFERENTIAL EQUATIONS

By suitably redefining variables it is always possible to bring a set of ordinary differential equations into a set of *first-order* differential equations of the type

$$\frac{\mathrm{d}}{\mathrm{d}t}\boldsymbol{x} = \boldsymbol{F}(\boldsymbol{x}, t, \mu). \tag{9.1}$$

The phase space spanned by the variables $\boldsymbol{x}$ is just a slight generalization of the concept of phase space used in elementary mechanics.

The parameters μ play a fundamental role since we shall often concentrate the attention on the *changes* in phase space as functions of the control parameters as we have seen already in the bifurcation pattern of the iterative maps.

In equation (9.1) an explicit dependence on time is allowed for. There are some basic differences between systems that have and those that do not have such an explicit time dependence. We shall only consider the latter, which are said to be *autonomous*.

Given a certain initial point in phase space, the time development of the system will trace out a *unique* phase space trajectory, and this trajectory will be the same independent of what time the system starts at that same initial point. It also follows from this that for autonomous systems, phase space trajectories cannot cross. (Note that this is absolutely not true for non-autonomous systems.) For this reason one-dimensional autonomous systems are trivial, and two-dimensional systems most likely will either converge to a point or 'explode', meaning that at least one of the variables will wander off to infinity. (Again, this is in general not true for non-autonomous systems.) In addition there is the possibility that the orbit is periodic, either unstable or stable. In the last case the orbit is attracting and is usually referred to as a *limit cycle*.

Consider several nearby initial points and their associated trajectories. Since they do not cross we may think of them as stream lines, a *flow*, in phase space.

We want to study how a small volume element evolves in time. To do this we consider the Jacobian matrix of the transformation $\boldsymbol{x}(0) \rightarrow \boldsymbol{x}(t)$. We then have the definition

$$J_{ik}(t) = \frac{\partial x_i(t)}{\partial x_k(0)}. \tag{9.2}$$

The corresponding Jacobian is called J. The local Jacobian matrix is defined to be

$$j_{ik}(t) = \frac{\partial F_i}{\partial x_k}. \tag{9.3}$$

Differentiating equation (9.2) with respect to time gives

$$\frac{d}{dt} J_{ik} = \frac{\partial F_i}{\partial x_k(0)} = \sum_{l=1}^{n} \frac{\partial F_i}{\partial x_l(t)} \frac{\partial x_l(t)}{\partial x_k(0)}$$

$$= \sum_{l=1}^{n} j_{il} J_{lk}. \tag{9.4}$$

Integrating this expression tells us exactly how a small volume element changes in time when it is carried along with the flow, and this includes changes in shape.

The magnitude of the volume element is given by

$$\delta V(t) = J(t)\,\delta V(0) \tag{9.5}$$

and if we just want to find the change in time of $\delta V(t)$, we have to find an expression for $dJ(t)/dt$ that can be integrated to find $J(t)$. The work is made easier using the symbolic notation

$$J = \frac{\partial(x_1(t), x_2(t), \ldots, x_n(t))}{\partial(x_1(0), x_2(0), \ldots, x_n(0))}. \tag{9.6}$$

We also need the chain rule for Jacobians

$$\frac{\partial(x_1, x_2, \ldots, x_n)}{\partial(z_1, z_2, \ldots, z_n)} = \frac{\partial(x_1, x_2, \ldots, x_n)}{\partial(y_1, y_2, \ldots, y_n)} \frac{\partial(y_1, y_2, \ldots, y_n)}{\partial(z_1, z_2, \ldots, z_n)}. \tag{9.7}$$

Differentiating J gives

$$\frac{d}{dt} J(t) = \frac{\partial(\dot{x}_1(t), x_2(t), \ldots, x_n(t))}{\partial(x_1(0), x_2(0), \ldots, x_n(0))} + \frac{\partial(x_1(t), \dot{x}_2(t), \ldots, x_n(t))}{\partial(x_1(0), x_2(0), \ldots, x_n(0))}$$

$$+ \cdots + \frac{\partial(x_1(t), x_2(t), \ldots, \dot{x}_n(t))}{\partial(x_1(0), x_2(0), \ldots, x_n(0))}. \tag{9.8}$$

Here the notation $dx(t)/dt = \dot{x}(t)$ has been used. Applying the chain rule to the ith term of the sum of equation (9.8) gives

$$\frac{\partial(x_1(t), x_2(t), \ldots, \dot{x}_i(t), \ldots, x_n(t))}{\partial(x_1(0), x_2(0), \ldots, x_n(0))}$$

$$= \frac{\partial(x_1(t), x_2(t), \ldots, \dot{x}_i(t), \ldots, x_n(t))}{\partial(x_1(t), x_2(t), \ldots, x_n(t))} \frac{\partial(x_1(t), x_2(t), \ldots, x_n(t))}{\partial(x_1(0), x_2(0), \ldots, x_n(0))}. \tag{9.9}$$

The last factor on the right-hand side of this equation is the Jacobian itself, and the first factor is

$$\frac{\partial(x_1(t), x_2(t), \ldots, \dot{x}_i(t), \ldots, x_n(t))}{\partial(x_1(t), x_2(t), \ldots, x_n(t))} = \frac{\partial \dot{x}_i(t)}{\partial x_i(t)} = \frac{\partial F_i(t)}{\partial x_i(t)}. \tag{9.10}$$

This finally gives

$$\frac{d}{dt} J(t) = \sum_{i=1}^{n} \frac{\partial F_i(x(t))}{\partial x_i} J(t) = (\text{Div } F) J(t) \tag{9.11}$$

and in the particular case that the divergence is a constant

$$J(t) \propto e^{(\text{Div } F)t}. \tag{9.12}$$

For a dissipative system this constant has to be negative, so there is an exponential contraction of phase space volume. If the divergence is not constant, it need not be negative everywhere. It suffices that it is negative in some 'average' sense. For a conservative system the divergence is zero which leaves phase space volumes invariant. (Of course the shapes are not invariant.)

9.1 Fixed points. Linear stability analysis

Assume that the system is such that there exist one or more fixed points x_0 in phase space where

$$F(x_0, \mu) = 0. \tag{9.13}$$

Linearizing around x_0 by putting $x = x_0 + \delta x$ and expanding the right-hand side of equation (9.1) to first order in δx gives

$$\delta \dot{x}_i = \sum_j \frac{\partial F_i(x = x_0)}{\partial x_j} \delta x_j(t). \tag{9.14}$$

The Jacobian matrix to the right is just a constant matrix and we shall assume that its elements are real and that the matrix may be diagonalized. Since the coefficients of the eigenvalue equation are real there can only be two types of eigenvalues, either real or complex conjugated sets.

Real negative eigenvalues correspond to an exponentially damped motion in the direction given by the eigenvector. In the case that *all* eigenvalues are real and negative, the fixed point is stable. This kind of stable fixed point attractor is called a stable *node*.

Real positive eigenvalues correspond to an accelerated, exponential motion away from the fixed point in the direction of the eigenvalue. In such a case the fixed point is unstable: it is a repellor. For a fixed point to be a repellor it is sufficient that one eigenvalue is positive. Many interesting models have fixed points where there are real parts of the eigenvalues of both signs, so that the orbits get injected into the region of the fixed point along some directions corresponding to negative real parts of the eigenvalues and ejected along directions corresponding to positive real parts of the eigenvalues. If there are no imaginary parts, the repellor is said to be an unstable node.

Complex conjugated eigenvalues correspond to a spiralling motion in a plane. Suppose one of these sets may be written

$$\lambda_{1,2} = \rho \pm i\,\omega. \tag{9.15}$$

Then the motion near x_0 is given by

$$\delta x(t) = c_1\,e^{\lambda_1 t}\,e_1 + c_2\,e^{\lambda_2 t}\,e_2 \tag{9.16}$$

where the eigenvectors of j are e_1 and e_2, and the two complex constants of the integration are so chosen that δx becomes a real vector, i.e. it becomes a sum of trigonometric functions times an exponential:

$$\delta x_i \sim e^{\rho t}\cos(\omega t + \text{phase}). \tag{9.17}$$

Evidently the motion is a spiral in the plane spanned by the two eigenvectors. The spiralling is inwards towards the fixed point when ρ is negative. In this case the fixed point is said to be a stable *focus*. When ρ is positive the spiralling is outwards and the focus is unstable.

In three dimensions there can either be three real eigenvalues or one real and a complex conjugated pair of eigenvalues. If the corresponding node or focus is unstable there are only two types of motion possible near the fixed point. The situation has been illustrated schematically in figures 9.1 and 9.2. Notice that there are no arrows on the orbits. This is because both directions are possible, depending on the signs of the real parts of the eigenvalues.

9.2 Homoclinic and heteroclinic orbits

We may define stable and unstable manifolds to be associated with the stable and unstable directions of the fixed points of flows just as we did for the multidimensional maps. For certain values of the parameters it may be possible to find initial points on the unstable manifold that are such that the orbit hits the stable manifold. Such an orbit will, after an infinite amount of time, hit the fixed point from which it once started out. This kind of orbit is called *homoclinic*. Correspondingly, *heteroclinic* orbits are

Node

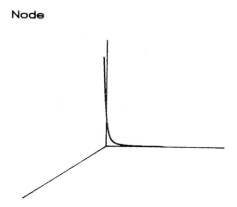

Figure 9.1 The motion near an unstable node in three dimensions.

Focus

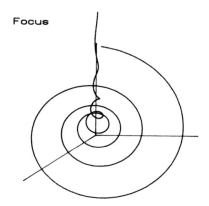

Figure 9.2 The motion near an unstable focus in three dimensions.

orbits that start out on the unstable manifold of one fixed point and end up on the stable manifold of another fixed point.

If the parameter is changed slightly and systematically in the vicinity of the parameter point where a homoclinic orbit has been found, there will normally be points where a type of periodic orbit can be found which is topologically similar to the homoclinic orbit. These orbits are only passing near the fixed point, and the system takes a finite time to move around the orbit. A periodic orbit that originates in this way is said to be born in a *homoclinic explosion*. If the fixed point is a node, the number of periodic orbits that can be produced in the explosion is small. For instance, if the unstable manifold is one dimensional there can be at most two: one for each possible direction away from the fixed point along the unstable

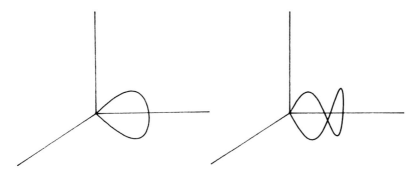

Figure 9.3 Schematic examples of homoclinic orbits to a node.

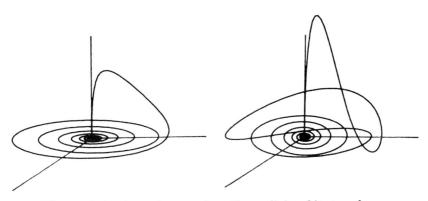

Figure 9.4 Schematic examples of homoclinic orbits to a focus.

manifold. However, if the fixed point is a focus there will be an infinity of
homoclinic orbits from the explosion. If there is one orbit that swings N
times around the fixed point, and N is a very big number, then close to it
in parameter space there will be another periodic orbit that swings $N - 1$
times around the fixed point and so on. This series may well terminate
at a number of swings greater than one, but at any rate there are an
infinity of periodic orbits originating in the explosion. Periodic orbits born
in homoclinic explosions may have no parameter range where they are
stable, while others may have some finite parameter interval of stability,
disconnected from the place of birth. For an example see chapter 10.

9.3 Lyapunov exponents for flows

In section 5.3 we learned that the Lyapunov exponents could be deduced from the eigenvalues of the Jacobian matrix of the map $x_0 \to x_T$. The complete analogue is true for flows: the Lyapunov exponents are deduced from the eigenvalues, $\Lambda_i(t)$, of the Jacobian of the map $x(0) \to x(t)$. The Lyapunov exponents are simply

$$\lambda_i = \lim_{t \to \infty} \frac{1}{t} \ln |\Lambda_i(t)|. \tag{9.18}$$

The Jacobian matrix may be found in principle by integrating equation (9.4), but as in the case of the discrete maps it is a non-trivial problem to calculate all but the largest LE. A simple algorithm for calculating the LEs of low-dimensional systems is given in appendix 2.

Without proof we state a few useful results concerning the LEs.

1. If the orbit of a dissipative system does not end on a fixed point, at least one of the LEs is zero.

2. For all initial points within one basin of attraction, the LEs are independent of the initial point.

3. If the divergence of the field, Div F, is a constant then

$$\sum_i \lambda_i = \text{Div } F. \tag{9.19}$$

In the special case of a conservative system, the divergence is zero. In that case there is the stronger version of 3.

4. For a conservative system the relation

$$\lambda_i + \lambda_{2N-i+1} = 0 \tag{9.20}$$

holds for all $0 < i < N + 1$, where N is the configuration space dimension (the phase space has dimension $2N$).

The LEs have a geometrical meaning. Corresponding to a negative LE there is a direction (in general not fixed) along the orbit where the orbit is attracting. Thus another nearby orbit in that direction will approach the original orbit. Exactly the opposite holds for directions corresponding to positive LEs. On the other hand, if a LE is zero it corresponds to some sort of 'frictionless' direction, normally in the direction of the orbit itself. Thus for a stable, periodic orbit (a limit cycle) all but one of the LEs are negative and one is zero. In general there may be scores of periodic orbits, but only very few of them are limit cycles.

9.4 Hopf bifurcations for flows

Although chaotic motion is not possible in flows of dimension less than three, the important phenomenon of a Hopf bifurcation may be illustrated by a two-dimensional example. Let the equations of motion be

$$\dot{r} = \mu r + \epsilon r^2$$
$$\dot{\theta} = -1. \qquad (9.21)$$

Here μ is a free parameter and $\epsilon = \pm 1$.

The system is trivially integrable. For $\epsilon = -1$ the result is

$$r = \frac{-\mu e^{\mu(t-t^0)}}{1 - e^{\mu(t-t^0)}} \quad \text{for } \mu < 0 \text{ or } \mu > 0 \text{ and } r > \mu$$

$$r = \frac{\mu e^{\mu(t-t^0)}}{1 + e^{\mu(t-t^0)}} \quad \text{for } \mu > 0 \text{ and } r < \mu$$

$$\theta = \theta_0 - t. \qquad (9.22)$$

For $\mu < 0$, $r \to 0$, so the origin is a stable point, and the general orbit is a spiral that ends at the origin.

For $\mu > 0$, $r \to \mu$ whether the orbit starts inside or outside the circle $r = \mu$ except if it starts at the origin. If it starts at the origin the orbit will forever remain there. Thus the origin is now an unstable fixed point, while the circle $r = \mu$ has become a limit cycle. This bifurcation of a stable point into a limit cycle is called a Hopf bifurcation.

The general Jacobian matrix is given by

$$j = \begin{pmatrix} \mu - 2r & 0 \\ 0 & 0 \end{pmatrix}. \qquad (9.23)$$

This matrix is already diagonal and its constant diagonal element must therefore be identical to one of the two Lyapunov exponents. This exponent is zero, in agreement with the general theorem cited in section 9.3. For the orbit consisting of the fixed point, the diagonal elements of j are μ and 0, showing what we already know: that the origin is stable for $\mu < 0$ and unstable for $\mu > 0$.

The orbit $r = \mu$ only exists for $\mu > 0$ and j has diagonal elements $-\mu$ and 0 which shows that it is a limit cycle.

Also, when $\epsilon = 1$ in equation (9.21) the integration is trivial and it shows that the motion is not necessarily bounded. For the fixed point the diagonal elements of j are still μ and 0, so the origin is a stable fixed point for $\mu < 0$. There is also the solution $r = -\mu$ which exists only for negative μ. The diagonal elements of the Jacobian are $-\mu$ and 0.

Since one of these is positive for the parameters where the orbit exists it follows that the periodic orbit is unstable. This phenomenon, where a stable fixed point and an unstable cyclic orbit at some point coalesce to form an unstable fixed point, is called an inverse Hopf bifurcation. In the Lorenz model (chapter 10) orbits born in homoclinic explosions at one fixed point disappear in inverse Hopf bifurcations at another fixed point.

10

THE LORENZ MODEL

The Lorenz model was originally constructed by truncating Fourier expansions of the Navier–Stokes equations. Somewhat unfortunately it turns out that this truncated system is a very bad approximation to the original equations except for the parameter range where the model is not very interesting. Nevertheless the model has become some sort of a numerical laboratory because it contains in a simple manner many of the fundamental features that nonlinear models may have. In particular it has regions of turbulence or deterministic chaos.

The equations of motion are given by

$$
\begin{aligned}
\dot{x} &= -\sigma(x - y) \\
\dot{y} &= (r - z)x - y \\
\dot{z} &= xy - bz
\end{aligned}
\tag{10.1}
$$

where x, y and z are real functions of time and σ, r and b are real, positive parameters. Most numerical experiments have been performed by fixing $\sigma = 10$ and $b = \frac{8}{3}$, and letting the 'Rayleigh number' r be the only parameter that is varied. We shall follow this practice.

The equations are seen to be invariant under reflection about the z axis $(x, y, z) \to (-x, -y, z)$. Notice that orbits in general do not possess this symmetry, but orbits found by the reflection transformation are also possible orbits.

The origin, O, and $C^{\pm} = (\pm\sqrt{b(r - 1)}, \pm\sqrt{b(r - 1)}, r - 1)$ are the fixed points. From this expression follows that $C^{\pm}$ only exist when $r > 1$. Linearizing equation (10.1) about these fixed points leads to the following eigenvalue equations for the Jacobian matrices of the fixed points:

$$
[\lambda^2 + (1 + \sigma\lambda) - \sigma(r - 1)](\lambda + b) = 0 \text{ for O} \tag{10.2}
$$
$$
\lambda^3 + (1 + \sigma + b)\lambda^2 + b(r + \sigma)\lambda + 2\sigma b(r - 1) = 0 \text{ for } C^{\pm}. \tag{10.3}
$$

For $0 < r < 1$ all eigenvalues are real and negative, so the origin is an attractor, and all orbits end up at the origin. At $r = 1$, the system goes through a pitchfork bifurcation whereby O loses its stability, while the two stable fixed points C^+ and C^- are created. Since O now has both a two-dimensional stable (associated with two different negative eigenvalues) and a one-dimensional unstable manifold one may start looking for homoclinic orbits to the origin. Starting out along the unstable manifold, the orbits always end on one of the three-dimensional stable manifolds of $C^\pm$ until $r = 13.56\ldots$ where the first homoclinic orbit is found. This homoclinic orbit makes just one swing around one of the points $C^\pm$ (which one is dependent on the direction taken along the unstable manifold of O) before it returns to the origin. The next homoclinic orbit is found at $r = 24.06\ldots$ and is of an entirely different nature since it encircles C^+ or C^- an infinity of times. However, we shall refrain from going into details of where in parameter space the homoclinic orbits to the origin occur.

Return to the first homoclinic orbit to see what happens when r is increased just a little bit. One may then find a periodic orbit that comes very close to the origin and therefore to being a homoclinic orbit. A periodic orbit which in this way is obviously related to a particular homoclinic orbit is often said to be *born* at the parameter point where that particular homoclinic orbit exists. Note that for an orbit that is not reflection symmetric, which the first homoclinic orbit is not, there always exists another with the same properties found by reflection through the z axis. As r is increased further each of the orbits born as homoclinic orbits encircles one of the points $C^\pm$ more and more closely, and finally at $r = 27.74\ldots$ disappears in an inverse Hopf bifurcation exactly at the r value where the points $C^\pm$ also lose stability. In the whole r interval in which they exist, these periodic orbits are unstable.

Let us return to the fixed points $C^\pm$ which were born in the pitchfork bifurcation at $r = 1$. A close study of equation (10.3) shows that there is one real eigenvalue of the Jacobian matrix, and in the parameter range of interest to us it is always negative, meaning that there will always be at least a one-dimensional stable manifold to $C^\pm$. The other two eigenvalues are complex conjugated with a negative real part for $1 < r < 24.74\ldots$. Thus, in this parameter range $C^\pm$ are stable foci, so the typical orbit is 'sucked' inwards along the direction corresponding to the direction of the stable manifold. At the same time the orbit spirals inwards around the focus in question approximately in the plane spanned by the eigenvectors corresponding to the complex conjugated set of eigenvectors.

The eigenvectors of the Jacobian matrix are of course only tangent vectors to the corresponding stable or unstable manifolds. When moving away from the fixed points the nonlinearity of the model becomes crucial and leads to an extremely complicated folding or curling of the manifolds, the shapes of which are important in 'understanding' why orbits behave the way they do.

Sticking to the simpler features of the model we now realize that for $r > 27.74\ldots$, any orbit that comes near the now unstable two-dimensional manifold of $C^\pm$ will have to spiral outwards because of the positive real part of the relevant eigenvalues of the Jacobians of $C^\pm$. The particular form of the nonlinearity will prevent the orbit from wandering off to infinity. Sooner or later, the orbit will come into the vicinity of the stable manifold of the

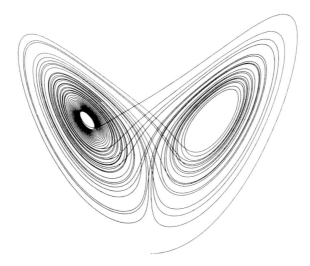

Figure 10.1 Orbit at $r = 28.0$. It is not possible to predict the number of revolutions around each focus without knowing the full system of equations and the initial conditions.

other fixed point, and thus gets attracted towards the other focus where it again spirals outwards a number of times before it is attracted to the first focus, and so on. There are no limit cycles in the region just above the inverse Hopf bifurcation, so phase space is completely dominated by chaotic orbits with sensitive dependence on initial conditions. A typical orbit is shown in figure 10.1.

From the equations of motion (10.1) it is easy to calculate the divergence. It is $-1 - b = -13.666\ldots$ which is a constant. From equation (9.12) this means that there is a phase space volume contraction by a factor $e^{-13.66t}$ which is very drastic. From the previous analysis we realize that there must be an expansion in the directions of a plane, so all the contraction must take place in one direction. The motion is said to be attracted to a strange attractor, the geometry of which is fractal and it is often referred to, perhaps not strictly correctly, as a *Cantor book*. At $r = 28$ numerical box-counting algorithms and the Kaplan–Yorke conjecture both give the same fractal dimension $d = 2.06$.

As r is increased there appear windows of stable periodic orbits. These windows are generally of the following nature: the simplest orbit (the parent or mother orbit) appears at the high-r end. If the orbit is itself invariant under the reflection symmetry of the system, the first bifurcation is a *symmetry-breaking* bifurcation whereby the topology of the orbit does not change, but the orbit is no longer symmetric. As r is decreased further in the window there follows a series of period doubling bifurcations precisely

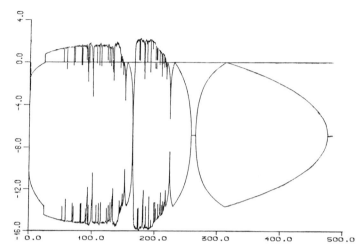

Figure 10.2 The three Lyapunov exponents in the range $0 < r < 500$. The 'line' at the value zero is actually a 'relay' between the largest and the second largest LE illustrating the theorem that if there is no stable fixed point, then one of the LEs is zero. For small r there *is* a stable fixed point. From Frøyland J and Alfsen K H 1984 *Phys. Rev.* A **29** 2928. Reproduced with permission.

as in the logistic map, and the constants of Feigenbaum are retrieved. The positions and widths of the windows can be read off from a compilation of Lyapunov exponents, such as the one shown in figure 10.2. In figure 10.4 a selection of symmetric orbits are shown. These are all mother orbits in their respective windows. Windows where the mother orbit is an asymmetric orbit also appear in great numbers. Figure 10.5 shows two heteroclinic mirror orbits and a homoclinic orbit. These orbits encircle the relevant fixed point an infinity of times. The orbits approach the fixed point along its stable one-dimensional manifold, and indeed the orbits are identical to the stable manifolds. On the other hand, the orbits start out in the plane of the unstable manifolds, and must therefore continue to stay in the two-dimensional unstable manifolds.

Consider a region very near the fixed point and any line from the fixed point in the plane of the unstable manifold. This line will be intersected by the homoclinic (heteroclinic) orbit at an infinity of points. In the linearized approximation the distances between these points form a geometric series. If we consider all points on the interval between two neighbouring intersection points, and the (unique) orbit going through each of these points, then the ensemble of these orbits must trace out the whole unstable manifold— an extremely curved and folded object, very difficult to picture.

Changing the parameters slightly, one can find closed orbits which do not follow the stable manifold exactly, and which only swing around the stable manifold a large, but finite number of times. The orbit is close

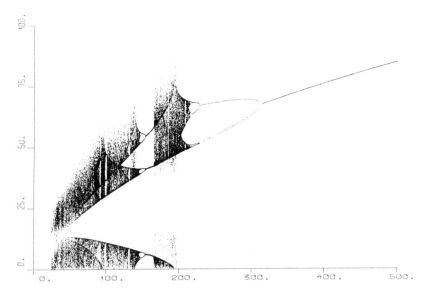

Figure 10.3 A Poincaré plot at $b = \frac{3}{8}$ in the range $24 < r < 520$. The distance to the origin is plotted each time the orbit intersects the plane $x = y$. Compare to figure 10.2. From Frøyland J and Alfsen K H 1984 *Phys. Rev.* A **29** 2928. Reproduced with permission.

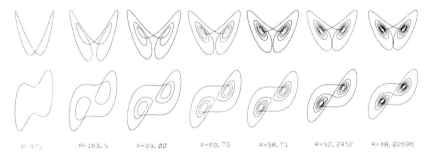

Figure 10.4 Projections into the xz plane (upper row) and the xy plane (lower row) of stable, symmetric orbits. From Frøyland J and Alfsen K H 1984 *Phys. Rev.* A **29** 2928. Reproduced with permission.

to, but not on, the unstable manifold surface. Going a little bit further away from the fixed point, one finds a similar orbit, but with one less swing around the stable direction. This process can be continued till the orbit does not perform any more swings. Orbits generated in this way are generally unstable, but they can be followed in parameter space, and many of them become stable in some window of parameter space. See figure 10.4.

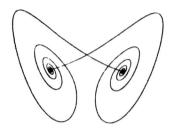
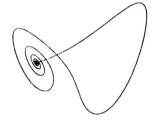

Figure 10.5 On the left the xy projections of two mirror heteroclinic orbits between C^+ and C^- at $\sigma = 10$, $b = 0.5$ and $r = 233.95$. On the right homoclinic orbit to C^- at $\sigma = 10$, $b = 0.5$ and $r = 233.99$. Between these two parameter points there exist homoclinic orbits to C^- that swing $2, 3, 4, \ldots$ times around C^+ before they hit C^-. Thus there are infinitely many homoclinic orbits in this extremely short interval in r. From Frøyland J and Alfsen K H 1985 *Physica Scr.* **31** 15. Reproduced with permission.

11

TIME SERIES ANALYSIS

Many experiments give as output long records of numbers produced by recording the observed value of some measurable quantity at regular time intervals. Such *time series* normally need to be analysed to be of more immediate use. There are many ways to do this. We shall not explain the more commonly used methods of analysis like power spectra analysis, frequency analysis, correlations, etc. Instead we shall be concerned with some aspects of time series analysis that are relevant to the preceding chapters.

11.1 Fractal dimension from a time series

We start by making some definitions: the (unobserved) n-dimensional phase state vector at some time t is $X(t)$ with components $\{x_0(t), x_1(t), \ldots, x_n(t)\}$. The time series is $\{x_0(t_1), x_0(t_2), \ldots, x_0(t_N)\}$ where we have not specified the size of the time intervals. We assume that each phase space variable is given by a first-order differential equation. By successive differentiations and eliminations of first-order terms in all but the variable $x_0(t)$ and its $n-1$ derivatives, the set of n differential equations in n variables may, at least formally, be transformed into a nth-order equation of just one variable. Having solved this equation, all the other variables are known. This is the basis for claiming that if one knows a time series at all times in one variable, everything is implicitly known about the system. In particular, a global quantity like the dimension of the attractor can be found from the time series of just one variable.

Before we proceed we notice that the boundary conditions may be fixed either by giving the function and each of its $n-1$ derivatives at one time, or alternatively giving the value of the function at n different times, thus avoiding having to deal with very high order derivatives.

We now define a new m-dimensional phase space using a time lag τ. A point in this new phase space is given by

$$\{x_0(t), x_0(t+\tau), x_0(t+2\tau), \ldots, x_0(t+(m-1)\tau)\}. \tag{11.1}$$

In vector notation we denote this point $x(t)$. As time runs, a trajectory is traced out in the new phase space, and this trajectory forms the basis of the algorithm for finding the dimension of a possible attractor embedded in the original phase space. The time lag τ is chosen so that the shifted variables can be expected to be independent or uncorrelated. Choosing τ large enough this condition will normally be fulfilled. This, however, does not mean that the choice of τ is irrelevant in practical applications, and we shall return to the question of how to choose τ shortly.

The actual observations are made at times $t_1, t_2, \ldots, t_N$ with corresponding phase space points $\boldsymbol{x}_1, \boldsymbol{x}_2, \ldots, \boldsymbol{x}_{N-m}$, and it is assumed that the system has been running sufficiently long for the transients to have died out before the measurements start. A standard euclidian distance measure may be used to define a distance $|\boldsymbol{x}_i - \boldsymbol{x}_j|$ between any two points $\boldsymbol{x}_i$ and $\boldsymbol{x}_j$. We may then form the quantity

$$E_m(r) = \frac{2}{(N-m)(N-m-1)} \sum_{i=1}^{N-m-1} \sum_{j=i+1}^{N-m} \theta(r - |\boldsymbol{x}_i - \boldsymbol{x}_j|) \quad (11.2)$$

where the step function $\theta(x) = 0$ for $x < 0$ and $\theta(x) = 1$ for $x > 0$. The function $E_m(r)$ is a measure of how many pairs of points can be found within a distance r from another.

In practical calculations it may save computer time first to find the distribution of distances $\rho(r)$ which gives the number of distances in the interval $(r, r + \Delta r)$ where Δr is some small quantity. If $N \rightarrow \infty$, and $\Delta r \rightarrow 0$ we see that $E_m(r) = \int_0^r \rho(r)\mathrm{d}r / \int_0^R \rho(r)\mathrm{d}r$, where R is the largest distance between any two vectors.

Suppose that all the points lie at regular intervals on a line. For small distances and large N we find $E_m(r) \propto r^1$ independent of m for small m. Similarly we find $E_m(r) \propto r^2$ for points on a regular, two-dimensional grid. It is then conjectured that for a more general case

$$E_m(r) \propto r^{D_m} \quad \text{for } r \text{ small} \quad (11.3)$$

where D_m is the dimension of the attractor. In principle one may therefore find D_m by finding the slope of the plot $\log(E_m(r))$ versus $\log(r)$.

Clearly there is some largest distance in the problem. That implies that the function $E_m(r)$ will level off and become a constant for large r. This region of r is of no use in this connection. Furthermore there is some smallest distance so that $\log(E_m(r)) \rightarrow -\infty$ for r less than some finite value. In practice we must consider distances that are small, but not so small that the statistics become too bad (see figure 11.1). Evidently it will not always be possible to find a well-defined dimension in this way, but assuming that one finds a well-defined D_m the procedure is to start with some low value for the *embedding dimension* m. If there really exists an attractor of dimension D in the system, and $D > m$, the result will be $D_m \approx m$. One then increases m by one unit at a time. When m is approaching D, one will see D_m level off and for $m > D$ one has $D_m \approx D$

(see figure 11.2). The dimension D arrived at in this manner is often called the *correlation dimension* or the *scaling dimension*.

To fix the time lag τ we need to know something about the timescale of the process. Actually there may be several timescales involved, but we are mainly interested in the shortest. One convenient way to define a timescale is through the *autocorrelation function*

$$C(\Delta t) = \frac{\langle x_0(t)x_0(t + \Delta t)\rangle}{\langle x_0(t)x_0(t)\rangle}. \tag{11.4}$$

The brackets are shorthand for time averages. The notation can be somewhat misleading since $\langle a(t)\rangle = (\sum_t^T a(t))/T$ does not depend on t. In equation (11.4) it has been assumed that the time average of $\langle x_0(t)\rangle = 0$. If this condition is not fulfilled it can be met by subtracting $\langle x_0(t)\rangle$ at all times. Evidently $C(0) = 1$, and $C(\Delta t) < 1$ for $\Delta t > 0$. If $x_0(t)$ is a random variable, the time average $\langle x_0(t)x_0(t+\Delta t)\rangle$ must be small due to cancelling terms. For small Δt in a continously varying system there are some correlations, so that $C(\Delta t)$ initially is a smoothly decreasing function of Δt. In many cases the initial peak may be well parametrized as $e^{(-\Delta t/\tau)}$. The value of τ is referred to as the *autocorrelation length*, and may be taken as the time lag of the first part of this section. However, this may be a little too small and perhaps the autocorrelation length should be considered to be some lower limit of the optimal time lag. Given the observed time series, it is fairly obvious that an optimal time lag exists because on one hand one wants to get as many data points as possible in the phase space, and hence the time lag should be as small as possible. On the other hand, the time lag must not be so small that unwanted dependences are introduced.

If one wants to make a model of the process that generated the time series, it may be very important to find the correlation dimension since it shows that one needs *at least* as many variables as the smallest integer larger than D.

11.2 Autoregressive models

We now assume that the time series is given by $\{x_1, x_2, \ldots, x_N\}$. We also assume that the sytem is so large, and the time series so short that it cannot possibly contain all the information about the system. Therefore a model of the system cannot realistically avoid containing unknown components. In the autoregressive models these unknown components are supposed to be given in the form of stochastic terms that are fed into the system at each time step. More specifically, the autoregressive model of order n, the

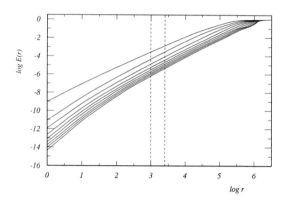

Figure 11.1 Plot of $\log(E_m(r))$ against $\log(r)$ for $m = 2, \ldots, 10$ from a time series taken at regular intervals of the Lorenz attractor.

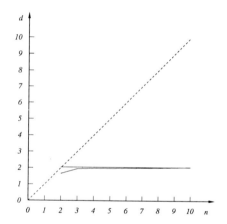

Figure 11.2 Computed attractor dimension D_m as a function of m for the Lorenz attractor. The horizontal line corresponds to the dimension 2.06 computed by other means.

$AR(n)$ model, makes the following assumption:

$$x_{t+1} = \sum_{i=1}^{n} a_i x_{t+1-i} + \sigma \xi_{t+1} \qquad (11.5)$$

where ξ_t is a random variable, usually considered to be a random Gaussian variable with zero mean and unit variance. The factor σ is a constant that regulates the strength of the stochastic term. Since the iterations are linear, the propagation of the effect of the stochastic term at one time is independent of all later or previous disturbances. Let k_i, $i = 1, 2, \ldots, n$, be the solutions of equation (5.7), and let $x_{t,0}$ be the result at time t of the disturbance $\sigma\xi_0$ at time zero. Then

$$x_{t,0} = \sum_{i=1}^{n} c_{i,0} k_i^t \tag{11.6}$$

where the coefficients $c_{i,0}$ are chosen so that all $x_{t,0}$ for negative times are zero:

$$\sum_{i=1}^{n} c_{i,0} k_i^t = 0 \quad \text{for } t = -(n-1), -(n-2), \ldots, -1$$

$$\sum_{i=1}^{n} c_{i,0} = \sigma\xi_0. \tag{11.7}$$

In a similar fashion one may determine coefficients $c_{i,t'}$ from the stochastic term at time t', so that one finally has

$$x_{t+1} = \sum_{i=1}^{n} \sum_{t'=0}^{t} c_{i,t'} k_i^{t-t'}. \tag{11.8}$$

Evidently for the model to not 'explode' we must have $|k_i| \leq 1$, but the equality means that the effect of a disturbance at some time is felt without damping at all later times. Therefore the equality also cannot be accepted in a model of a dissipative system.

From just looking at the numbers a_i it is usually not so easy to get a good impression of the content of the model. A simple and instructive way to get a geometrical picture of how the model works can be obtained by using the iterative map

$$R_{t+1} = \sum_{i=1}^{n} a_i R_{t+1-i} + \delta_{t+1,0} \tag{11.9}$$

with the initial contitions $R_t = 0$ for all $t < 0$. Here $\delta_{i,i} = 1$ and $\delta_{i,j} = 0$ for $i \neq j$. We shall call the function R_t the *response function* of the system or model. Given the stochastic terms of equation (11.5) and R_t at all times prior to some time, all results can be expressed in terms of these quantities. This is because the response of the system to any disturbance is linear so that the effect of each disturbance $\sigma\xi_t$ at any later time can be calculated without taking into account any other disturbance. Thus the total result can be written

$$x_t = \sigma \sum_{t'=0}^{t} R_{t'} \xi_{t-t'}. \tag{11.10}$$

Here we set the boundary condition $\xi_0 = x_0$. Given the coefficients a_i it is always trivial to find R_t from equation (11.9) by straightforward iterations with initial conditions as described above. Evidently R_t can be expressed as a sum of exponentials of the form $R_t = \sum_i^n r_i k_i^t$ where the coefficients r_i can be determined in a way completely similar to the way $c_{i,0}$ was determined.

In practice we have to find the coefficients a_i from the time series itself, and we may do this by choosing the coefficients so that the sum of the squares of the necessary stochastic terms becomes as small as possible. For a given time series we then determine not only the coefficients a_i, but also exactly which stochastic terms are necessary to exactly reproduce the time series. Accordingly we minimize the function

$$Q_n = \sum_{t=n}^{N} \left(x_t - \sum_{i=1}^{n} a_i x_{t-i} \right)^2. \tag{11.11}$$

Differentiating with respect to each a_j and setting the result equal to zero one finds

$$\sum_{i=1}^{n} a_i \left(\sum_{t=n}^{N} x_{t-i} x_{t-j} \right) = \sum_{t=n}^{N} x_t x_{t-j} \text{ for } j = 1, \ldots, n. \tag{11.12}$$

These are the *Yule–Walker equations* to determine the coefficients a_i. If we divide by $\sum_t (x_t)^2$ everywhere we see that what is inside the parenthesis on the left-hand side of equation (11.12) is nothing but the autocorrelation function $C(i - j)$ which we notice is symmetric in the indices i and j. On the right-hand side we have the autocorrelation function $C(j)$. As a result equation (11.12) may be rewritten

$$\sum_{i=1}^{n} a_i C(i - j) = C(j) \text{ for } j = 1, \ldots, n. \tag{11.13}$$

So far we have assumed the order, n, of the regression to have been determined in advance. This will rarely be the case, so we have to find a way of fixing n. The equations (11.13) are normally easy to solve by some suitable computer program. The result may then be inserted into the definition of Q_n. Increasing n starting at some low value leads to a monotonically falling Q_n. However, at some point the improvement becomes insignificant, and there is no point in increasing n any further.

Having determined a_i we may now use equation (11.5) to determine the stochastic part $\sigma \xi_t$ for $t = n, \ldots, N$ (we cannot find this term for $t < n$ because that would require knowledge of x_t at negative times), and subject it to tests to see if it fulfils the assumption of being a random Gaussian distribution. In particular we see that $\sigma = \sqrt{Q_n/(N - n)}$.

If the stochastic term is sufficiently close to being random and Gaussian, equation (11.5) may be used for *statistical forecasting*. We then consider the coefficients a_i and σ to be fixed, and use computer generated random

Gaussian distributed numbers for ξ_t to make the iterations for arbitrarily long times. This may then be used to compute various probabilities like the average largest and smallest excursions in some given time span, the number of expected events outside or inside some interval, etc.

Before we leave this very brief discussion of the $AR(n)$ model, it may be of value to recognize that fixing its coefficients a_i implicitly also fixes n timescales. For a real root k_i of equation (5.7) the timescale may be considered to be a decay or damping time τ defined through $|k_i|^t = e^{-t/\tau_i}$, so that $\tau_i = -1/\ln(|k_i|)$ which approaches infinity as $|k_i| \to 1$, and zero as $|k_i| \to 0$. Complex roots appear in pairs for real a_i, and any complex root may be written as $k_i = e^{(-1/\tau_i + i\omega_i)}$. Consequently a complex pair of roots localizes a damping time and a cyclic timescale $T_i = 2\pi/\omega_i$.

11.3 Rescaled range analysis

In the previous section we saw that the $AR(n)$ models in general had an exponentially decaying memory as expressed by the components of the response function. It may be that a given record has a memory of a different kind, for instance it could have a power law behaviour. This could never be detected by an autoregressive analysis. The signal that an $AR(n)$ model is unsatisfactory would be that the deduced record ξ_i would not fulfil the requirements of being an independent Gaussian process. However, long time memory in the record could be revealed by what is known as a *rescaled range analysis* and which will be briefly presented below.

The $AR(1)$ process

$$x_{t+1} = x_t + \xi_{t+1} \tag{11.14}$$

with ξ_t drawn from a Gaussian distribution of zero mean and unit variance is an independent Gaussian process known as a *random walk process*. After t time steps, the cumulative result is

$$x_t = \sum_{t'=0}^{t} \xi_{t'}. \tag{11.15}$$

If we assume that this process is repeated for an infinite number of times, each time starting at the origin, the distribution of results after t time steps will be another Gaussian with zero mean, but not unit variance. Denoting averages over attempts by brackets we have $\langle \xi_t \rangle = 0$ for all t and consequently also $\langle x_t \rangle = 0$ for all t. Evidently $\langle \xi_t \xi_{t'} \rangle = \delta_{t,t'}$. It follows that

$$\langle x_t^2 \rangle = \sum_{i,j=0}^{t} \xi_i \xi_j = t. \tag{11.16}$$

This means that the width of the distribution grows like $t^{1/2}$ with time.

Suppose we consider a random walk process for a length of time τ, and draw the steps of a new random walk process from the distribution of x_τ. After t steps the width of the new distribution is $(\tau t)^{1/2}$, or the same as it would have been with the original step lengths taken for τt steps. Evidently we could have obtained a valid distribution for a time τt from the distribution at time t by scaling all terms a factor $\tau^{1/2}$. For a more general process this generalizes into τ^H where the exponent is called the Hurst exponent. Equation (11.16) then becomes

$$\langle (x_t - \overline{x})^2 \rangle \sim t^{2H}. \tag{11.17}$$

Given a record it is not so straightforward to check if it has the scaling properties described above because it is a scaling property of distributions and not of the record itself. A method has been developed to find the Hurst exponent, and it is outlined in the following without proof.

Let the record be $\{x_t\}$ for $0 < t < T$. For a given time t and a given timelag τ we may define local averages

$$\overline{x}_{t,\tau} = \frac{1}{\tau} \sum_{l=t}^{t+\tau-1} x_l. \tag{11.18}$$

The local cumulative records

$$Y_{t,\tau,s} = \sum_{l=t}^{t+s-1} (x_l - \overline{x}_{t,\tau}) \qquad \text{for } 0 < s \le \tau \tag{11.19}$$

have the property $Y_{t,\tau,\tau} = 0$. When t, and τ are fixed, $Y_{t,\tau,s}$ as a function of s starts out at some small value, makes some excursions and ends at zero. This allows one to make a sensible definition of the *range* of the record. The local range is given by

$$R_{t,\tau} = \max(Y_{t,\tau,s}) - \min(Y_{t,\tau,s}) \tag{11.20}$$

where $\max/\min(Y_{t,\tau,s})$ means the largest/smallest value of $Y_{t,\tau,s}$ in the range $0 < s < \tau$. The standard deviations of the distributions are given by

$$S_{t,\tau}^2 = \frac{1}{\tau} \sum_{s=1}^{\tau} Y_{t,\tau,s}^2. \tag{11.21}$$

In order to make the procedure more robust, and to make it possible to compare different processes, the range is *rescaled* by the standard deviation. This means that for every time lag τ, and every time $t \in [0, T - \tau]$ we have a rescaled range quantity $R_{t,\tau}/S_{t,\tau}$. There is a lot of redundancy in these quantities. The redundancies can be reduced in various ways. The simplest is probably to divide the whole interval into n_τ subintervals of length τ so that $n_\tau \tau \leq T \leq (n_\tau + 1)\tau$ and use as only t values $t = i\tau$ where $0 < i < n_\tau$ and i is an integer, and then compute an average rescaled range for each τ value:

$$r_\tau = \frac{1}{n_\tau} \sum_{i=1}^{n_\tau} R_{i\tau,\tau}/S_{i\tau,\tau}. \tag{11.22}$$

Since r_τ can be computed directly from the original record, one may make a plot of $\ln(r_\tau)$ against $\ln(\tau)$. A straight line allows the Hurst exponent for the record to be determined. If a sensible Hurst exponent can be defined, $r_\tau \propto \tau^H$ for all scales or timelags τ within some reasonable range.

When $\frac{1}{2} < H < 1$ the record is said to show a *persistent* behaviour. When $0 < H < \frac{1}{2}$ the record is said to be *antipersistent*. When the record is persistent, a large excursion in one time interval is more likely to be followed by another large excursion in the next time interval of the same length, and this is true for *all* relevant timescales, including the longest. The autocorrelation function calculation could not reveal this type of long time behaviour.

For the above procedure to work, it is important that the record does not contain any cyclic components. In many cases such components can be removed quite easily.

11.4 The global temperature: an example

In figure 11.3 is shown a reconstructed temperature record of the average temperature of the atmosphere of the northern hemisphere for about 400 years. We subject the record to an analysis using an $AR(n)$ model of section 11.2. It turns out that there is no significant reduction of the stochastic terms for $n > 4$. The resulting stochastic term record for $n = 4$ is shown in figure 11.4. As one may see by inspection, the variance is much smaller than in the original data, and correlations in time that are very obvious in the original data are almost absent. In particular, the visible long time trends seem very faint.

As may be seen from table 11.1 there are two real roots of equation (5.7), and a complex conjugated pair of solutions. This corresponds to three damping times, and one cyclic time. It should be noted that an AR(2) model gives no cyclic time, and the difference between Q_2 and Q_4 is small, hence the calculated frequency is not necessarily very significant. The response function is displayed in figure 11.5, and as can be seen it falls quickly, but then has a fairly long tail so that there is still a noticeable effect after 15–20 years. We can also make a rescaled range analysis of the record and find $H = 0.92 \pm 0.05$ (see figure 11.6).

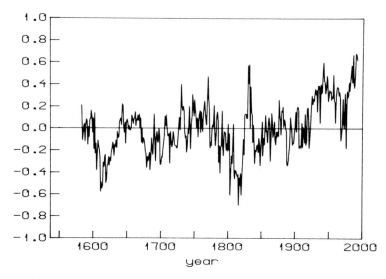

Figure 11.3 Reconstructed record of the annual average temperatures for the northern hemisphere during the last four centuries. The figure shows the deviations from the record mean in °C. The record mean is 13.8 °C. Record up to and including the year 1975 from Groveman B S and Landsberg H E 1979 *Meteorol. Prog., Univ. Maryland,* Publ 79-181 and Borzenkova I I *et al* 1976 *Meteorol. i Gidrol.* **7** 27. Data from the period 1976–1989 have been read off from figure 11 in *Climate Change—The IPCC Scientific Assessment* ed J T Houghton *et al* (Cambridge: Cambridge University Press) 1990. The data last referred to are not included in the analysis carried out in the text.

Table 11.1 Table of the coefficients of the AR(4) model using the reconstructed record of the annual average temperature on the northern hemisphere in figure 11.3. The roots of equation (5.7) are also given, together with the corresponding timescales.

i	1	2	3	4
a_i	0.5357	0.1595	-0.0165	0.1859
k_i	0.9250	-0.6063	$0.109+i0.5655$	$0.109-i0.5655$
τ_i (years)	12.7	2.00	1.81	
T (years)				4.55

From simple calculations of the autocorrelation function it is clear that there are strong correlations for times up to 8–10 years, but what the large Hurst exponent reveals is that there are correlations on much longer

Reduced data

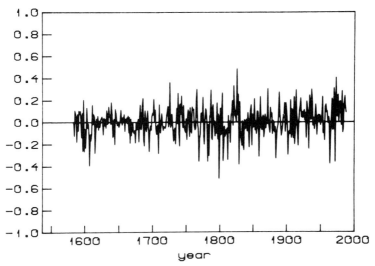

Figure 11.4 The necessary stochastic terms to produce the record in figure 11.3 in an AR(4) model.

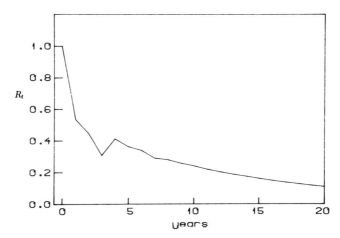

Figure 11.5 The response function of the AR(4) model resulting from the global temperature record of figure 11.3.

timescales, and that there is the same type of correlation on timescales from three years and up to about one hundred years. One can interpret the result in the following way: a cold year is most likely followed by another cold year, a cold decade is most likely followed by another cold decade and so on.

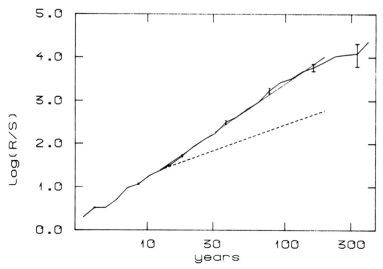

Figure 11.6 Logarithm of the average rescaled range plotted against logarithm of time lag.

We can also make a rescaled range analysis of the record ξ_i shown in figure 11.3. The result is $H_\xi = 0.58 \pm 0.04$ which shows that this record does not quite fulfil the requirements necessary in order for it to be an independent Gaussian process, which one must assume if one is to use the model for statistical forecasting. Therefore statistical forecasting based on the AR(4) model cannot be fully trusted.

The response function shown in figure 11.5 shows that there is a memory for 10–15 years. If the AR(4) model had been the whole truth, the R/S analysis would have produced a bent curve in figure 11.6 starting out steeply for times up to about 10 years, but which for times longer than 15–20 years would approach a line with a slope of $\frac{1}{2}$. There is no sign of such a bending except for times of the order of 100 years, but this may very well be due to the lack of data.

APPENDIX 1

PERIOD THREE IN THE LOGISTIC MAP

We assume that the map is in the form of equation (3.13), and in order to avoid indices we use (x, y, z) instead of (x_1, x_2, x_3). With these conventions the set of equations to solve becomes

$$\begin{aligned} y &= c + x^2 \\ z &= c + y^2 \\ x &= c + z^2. \end{aligned} \tag{A1.1}$$

It is convenient to introduce the quantity

$$s \equiv x + y + z. \tag{A1.2}$$

Summing the three equations (A1.1) we find

$$x^2 + y^2 + z^2 = s - 3c. \tag{A1.3}$$

On the other hand,

$$\begin{aligned} s^2 &= x^2 + y^2 + z^2 + 2(xy + yz + zx) \\ &= s - 3c + 2(xy + yz + zx). \end{aligned} \tag{A1.4}$$

Subtracting two of the equations (A1.1) in all three possible ways and factorizing gives

$$\begin{aligned} y - z &= (x + y)(x - y) \\ z - x &= (y + z)(y - z) \\ x - y &= (z + x)(z - x). \end{aligned} \tag{A1.5}$$

There are two obvious solutions to equation (A1.1), namely the two period one orbits obtained by putting $x = y = z$. At this point we are not interested in these solutions. Consequently we may divide away the common factor obtained by multiplying together the three equations (A1.5). This gives

$$\begin{aligned} 1 &= (x + y)(y + z)(z + x) \tag{A1.6} \\ &= (s - z)(s - x)(s - y) \tag{A1.7} \end{aligned}$$

Starting from the first of these equations one finds after using equations (A1.1)–(A1.4)

$$
\begin{aligned}
1 &= x^2(y+z) + y^2(z+x) + z^2(x+y) + 2xyz \\
&= (y-c)(y+z) + (z-c)(z+x) + (x-c)(x+y) + 2xyz \\
&= y^2 + x^2 + z^2 - 2c(x+y+z) + xy + yz + zx + 2xyz \\
&= \tfrac{1}{2}s^2 + (\tfrac{1}{2} - 2c)s - \tfrac{3}{2}c + 2xyz.
\end{aligned} \tag{A1.8}
$$

The second of the equations (A1.7) gives

$$
\begin{aligned}
1 &= s^3 - (x+y+z)s^2 + (xy+xz+yz)s - xyz \\
&= \tfrac{1}{2}(s^2 - s + 3c)s - xyz.
\end{aligned} \tag{A1.9}
$$

Eliminating the term xyz from equations (A1.8) and (A1.9) gives an equation that contains s as its only variable:

$$
\begin{aligned}
s^3 - \tfrac{1}{2}s^2 &+ (c + \tfrac{1}{2})s - 3(\tfrac{1}{2}c + 1) \\
&= (s - \tfrac{2}{3})(s^2 + s + c + 2) = 0.
\end{aligned} \tag{A1.10}
$$

The solutions are

$$
s_{1,2} = -\tfrac{1}{2} \pm \sqrt{-\tfrac{7}{4} - c}
$$

and

$$
s_3 = \tfrac{2}{3}. \tag{A1.11}
$$

The orbit visits three points, and we assume that these points are the roots of a third degree equation. Such an equation may be found by using the identity

$$
x^3 - (x+y+z)x^2 + (xy+yz+zx)x + xyz = 0. \tag{A1.12}
$$

When s is fixed the terms inside brackets and the xyz term are given by the definition (A1.2), equation (A1.4) and equation (A1.9). Putting $s = s_3$ does not correspond to any real orbit, and so we are left with the two possibilities $s = s_{1,2}$. Obviously we must have $c \leq -\tfrac{7}{4}$ for s to be real, but that does not guarantee real orbits. We see from equation (A1.11) that for $c = -\tfrac{7}{4}$ the two orbits coalesce. We know from section 3.13 about tangent bifurcations that the period three orbits are born in a tangent bifurcation that creates one stable and one unstable period three orbit. To check algebraically that two real orbits really exist involves some trivial algebra that will not be given here. The important point is that they exist.

For these orbits to be stable, we must have

$$
-1 < f'(x)f'(y)f'(z) = 8xyz < 1. \tag{A1.13}
$$

In these inequalities, let us denote either one of the values ± 1 by w, and exchange the two inequality signs in relations (A1.13) by equals signs. $w = 1$ corresponds to the tangent bifurcation, while $w = -1$ corresponds to the period doubling bifurcation. Using equations (A1.9) and (A1.11) the result may be expressed as

$$\pm\, 8c\sqrt{-\tfrac{7}{4} - c} = 4c + 8 - w. \qquad (A1.14)$$

In general this is equivalent to a third degree equation with restrictions on its solutions. For $w = 1$ one sees that $c = -\tfrac{7}{4}$ is indeed the solution. For $w = -1$ the equation (A1.14) with the negative sign to the left has one solution which is one of the roots of the third degree equation obtained by squaring equation (A1.14). This is the period doubling point of the stable period three orbit. The result is somewhat clumsy, and it is left for the reader as an exercise to find it. Equation (A1.14) with the positive sign can never be satisfied for any real c.

If we want to find the tangent bifurcation point expressed in terms of the parameter λ used in chapter 3 on the logistic map, we have to use the connection $c = 2\lambda(1 - 2\lambda)$, and this gives the result $\lambda_c = (1 + 2\sqrt{2})/4$ quoted in section 3.11.

APPENDIX 2

LYAPUNOV EXPONENTS ALGORITHM

There are several algorithms for calculating the Lyapunov exponents. The one that will be given here without proof is convenient for systems of dimensions less than four.

A2.1 Lyapunov exponents for maps

Assume that the map has the form

$$x_{t+1}^i = f^i(x_t^1, x_t^2, \ldots, x_t^n) \quad i = 1, \ldots, n \qquad (A2.1)$$

and that the Jacobian matrix is given by

$$j_{ik}(x_t) = \frac{\partial f^i}{\partial x^k}. \qquad (A2.2)$$

To compute the rth Lyapunov exponent, consider the components of an $\binom{n}{r}$-dimensional vector v_{t+1} given by the recursive relation

$$v_{t+1}^i = \sum_{k=1}^{\binom{n}{r}} d_{ik}^{(r)}(j) v_t^k. \qquad (A2.3)$$

Here $d^{(r)}(j)$ is a $\binom{n}{r} \times \binom{n}{r}$ matrix constructed from the Jacobian matrix by crossing out $(n-r)$ rows and $(n-r)$ columns in all possible ways. For each combination, the determinant of the remaining elements of j is a matrix element of $d^{(r)}$. The technical name of this matrix is the r-compound matrix of j, or the rth exterior power of j. For our purposes the matrices $d^{(k)}$ may be transformed by letting any row change sign as long as the sign is also changed for the column with the same index. Also, any two rows may be interchanged provided also the two columns with the same indices are interchanged.

In particular we notice that $d^{(1)} = j$, and $d^{(n)} = \text{Det } j$. Let $v_0^k = 1$ for all k. In all but a few special cases the answer will be the same if only one of the components of v_0 is different from zero. With these initial conditions, the vector v_t is computed along with the trajectory up to some problem-dependent, long time T. The sum of the r biggest Lyapunov exponents is then given by

$$\sum_{i=1}^{r} \lambda_i = \lim_{T \to \infty} \ln(\max|v_T^k|)/T \qquad (A2.4)$$

where $\max|v_T^k|$ is the component of v_T with the biggest absolute value. Thus, using $r = 1$, i.e. $d^{(r)} = j$, one finds the biggest Lyapunov exponent. For two-dimensional maps $d^{(2)} = \text{Det } j$, and using equation (A2.3) and (A2.4) one can find $\lambda_1 + \lambda_2$.

In the three-dimensional case $d^{(2)}$ is the 3×3 matrix

$$d^{(2)} = \begin{pmatrix} \begin{vmatrix} j_{11} & j_{12} \\ j_{21} & j_{22} \end{vmatrix} & \begin{vmatrix} j_{11} & j_{13} \\ j_{21} & j_{23} \end{vmatrix} & \begin{vmatrix} j_{12} & j_{13} \\ j_{22} & j_{23} \end{vmatrix} \\ \begin{vmatrix} j_{11} & j_{12} \\ j_{31} & j_{32} \end{vmatrix} & \begin{vmatrix} j_{11} & j_{13} \\ j_{31} & j_{33} \end{vmatrix} & \begin{vmatrix} j_{12} & j_{13} \\ j_{32} & j_{33} \end{vmatrix} \\ \begin{vmatrix} j_{21} & j_{22} \\ j_{31} & j_{32} \end{vmatrix} & \begin{vmatrix} j_{21} & j_{23} \\ j_{31} & j_{33} \end{vmatrix} & \begin{vmatrix} j_{22} & j_{23} \\ j_{32} & j_{33} \end{vmatrix} \end{pmatrix}. \qquad (A2.5)$$

Using this equation and equations (A2.3) and (A2.4) gives λ_1 and λ_2. Finally λ_3 is found using $d^{(3)} = \text{Det } j$.

A2.2 Lyapunov exponents for flows

Assume that the equations of motion are given by the autonomous system of ordinary differential equations

$$\dot{x}^i(t) = f^i(x^1(t), x^2(t), \ldots, x^n(t)) \qquad i = 1, \ldots, n. \qquad (A2.6)$$

The Jacobian matrix is still given by equation (A2.2), and as before we consider the time development of n $\binom{n}{r}$-dimensional vectors $v(t)$, $r = (1, 2, \ldots, n)$, which develop in time according to the set of differential equations

$$\dot{v}^i(t) = \sum_{k=1}^{\binom{n}{r}} M_{ik}^{(r)} v^k(t). \qquad (A2.7)$$

The $\binom{n}{r} \times \binom{n}{r}$ matrices $M^{(k)}$ are defined as

$$M^{(k)} = \lim_{\Delta t \to 0} [\{d^{(k)}(1 + j\Delta t) - 1\}/\Delta t]. \qquad (A2.8)$$

For $n < 4$ these matrices are very simple to express in terms of the componets of j, and are given explicitly below. For $r = 1$ we have $M^{(1)} = j$, and for $r = n$ we have

$$M^{(n)} = \text{Tr } j = \sum_{i=1}^{n} \frac{\partial f^i}{\partial x^i}. \qquad (A2.9)$$

If this divergence is a constant, equation (A2.7) for $r = n$ may be integrated directly to give

$$\sum_{i=1}^{n} \lambda_i = \sum_{i=1}^{n} \frac{\partial f^i}{\partial x^i}. \tag{A2.10}$$

For $r = n - 1$ one has

$$M_{lm}^{(n-1)} = \left(\sum_{i=1}^{n} \frac{\partial f^i}{\partial x^i} \right) \delta_{lm} - j_{ml}. \tag{A2.11}$$

In the case $n = 3$ we have explicitly

$$M^{(2)} = \begin{pmatrix} (j_{22} + j_{33}) & -j_{21} & -j_{31} \\ -j_{12} & (j_{33} + j_{11}) & -j_{32} \\ -j_{13} & -j_{23} & (j_{11} + j_{22}) \end{pmatrix}. \tag{A2.12}$$

Integrating the orbit and equation (A2.7) simultaneously up to some long time, T, the sums of the Lyapunov exponents can be found just as before from equation (A2.4).

As an example we have for the Lorenz model of chapter 10

$$M^{(1)} = \begin{pmatrix} -\sigma & \sigma & 0 \\ -z + r & -1 & -x \\ y & x & -b \end{pmatrix} \tag{A2.13}$$

$$M^{(2)} = \begin{pmatrix} -1 - b & z - r & -y \\ -\sigma & -\sigma - b & -x \\ 0 & x & -1 - \sigma \end{pmatrix} \tag{A2.14}$$

and

$$M^{(3)} = -1 - b - \sigma. \tag{A2.15}$$

Since $M^{(3)}$ is a constant it follows from the result of section 9.3 that

$$\lambda_1 + \lambda_2 + \lambda_3 = -1 - b - \sigma. \tag{A2.16}$$

Furthermore, one of the exponents λ_1 or λ_2 must be zero depending on whether the orbit is stable periodic or chaotic. This may be used as a check on the precision of the calculation.

A2.3 Practical hints

Except in very special cases the biggest component of the vectors v will either grow or decrease exponentially, and in computer calculations this will sooner or later lead to an overflow/underflow problem that has to be dealt with. This may be done by testing on the various components of the vectors v. Since the equations of motion for the vs are linear, they may be scaled arbitrarily at any time. We want to compute the logarithm of the components of v and therefore only need to remember the sum of the logarithms of the scaling factors. However, tests are expensive in computer time, so it is better to make the tests outside some inner loop with no tests, say every 20 or so time steps.

FURTHER READING

Collet, P. and Eckmann, J.-P.
Iterated Maps on the Interval as Dynamical Systems
Birkhäuser, Boston 1980

Cvitanovic, P.
Universality in Chaos 2nd edn
Adam Hilger, Bristol 1989

Devaney, R.L.
An Introduction to Chaotic Dynamical Systems
Benjamin/Cummings, Menlo Park 1986

Feder, J.
Fractals
Plenum Press, New York 1988

Gallavotti, G. and Zweifel, P.F.
Nonlinear Evolution and Chaotic Phenomena
Plenum Press, New York 1988

Guckenheimer, J., Moser, J., Newhouse, S.
Dynamical Systems
Birkhäuser, Boston 1980

Haken, H. ed.
Chaos and Order in Nature
Springer Series in Synergetics, 11, Springer-Verlag, Berlin 1981

Hao, B.-l.
Chaos II
World Scientific, Singapore 1990

Holden, Arun
Chaos
Manchester University Press, Manchester 1986

Lichtenberg, A. L. and Lieberman, M. A.
Regular and Stochastic Motion
Springer-Verlag, Berlin 1983

Mandelbrot, B. B.
Fractals: Form, Chance and Dimension
W.H.Freeman and Co., San Francisco 1977

Mandelbrot, B. B.
The Fractal Geometry of Nature
W.H.Freeman and Co., San Francisco 1983

Mira, Ch.
Chaotic Dynamics
World Scientific, Singapore 1987

Ozorio de Almeida, M.
Hamiltonian Systems—Chaos and Quantization
Cambridge University Press, Cambridge 1988

Peitgen H.-O. and Richter P.H.
The Beauty of Fractals
Springer-Verlag, Berlin 1986

Schuster, H. G.
Deterministic Chaos
Physik-Verlag GmbH, Weinheim 1984

Sparow, C.
The Lorenz Equations, Bufurcations, Chaos and Strange Attractors
Appl. Math. Sci., 38, Springer-Verlag, Berlin 1982

Wolfram, S.
Theory and Applications of Cellular Automata
World Scientific, Singapore 1986

INDEX